IMAGES
of America

NEIHART MINING

On the Cover: The photograph on the cover depicts almost everything that takes place around an operating mine. The key word here is "operating," meaning that the initial hardrock exploratory work is complete (on the surface, at least), and what is now visible represents the fruits of those labors. Horse-drawn ore cars; timbers stacked and ready for new bracing, cribbing, or steam power; and miners and laborers are visible in front of the snow house (which provides cover at the mine's entrance, or adit) and a structure most likely serving as an office and occasional assay laboratory. (Louis W. Hill Papers. MN Historical Society, Saint Paul.)

IMAGES
of America

NEIHART MINING

R.L. Lansverk

ARCADIA
PUBLISHING

ISBN 978-0-7385-9691-4

Published by Arcadia Publishing
Charleston, South Carolina

Printed in the United States of America

Library of Congress Control Number: 2013930566

For all general information, please contact Arcadia Publishing:
Telephone 843-853-2070
Fax 843-853-0044
E-mail sales@arcadiapublishing.com
For customer service and orders:
Toll-Free 1-888-313-2665

Visit us on the Internet at www.arcadiapublishing.com

To my mother, Betty Freeh Lansverk, and my father, Owen C. Lansverk. Having both of you by my side through all the years has allowed me to trust my instincts and continue exploring, learning, and sharing. Thank you.

Contents

ACKNOWLEDGMENTS

Writing a book, any book on any topic, becomes an adventure. You need experienced guidance from many and wisdom from a few. Fortunately, I found all of that and then some.

I believe it dangerous to extend acknowledgment to those who helped me on this journey, being fearful I might miss somebody. Yet, with that said, I present my gratefulness to the following individuals and organizations. I thank Judy Ellinghausen, formerly of the History Museum for Cascade County and Great Falls, Montana, for a decade of support for this project. She continued to offer encouragement, ideas, and the final push to get me over the edge. Kristen Bokovoy slid into Judy's shoes as if she had been working on this project with me for at least 10 years.

The History Museum for Cascade County and Great Falls showed great support when I decided a book of this sort was required. Mitchell W. Reynolds, PhD, Scientist Emeritus, US Geological Survey, helped me broaden my narrow grasp of geology to the point that I may have picked up another hobby. Lory Morrow, of the Montana Historical Society, offered patient assistance as I bugged and pleaded with her to find more and more historic photographs of Neihart, even when none existed. Longtime friend and photograph editor for *Montana the Magazine of Western History* Glenda Bradshaw had the right words to keep me persevering when I felt low from difficulty locating certain images. A huge debt of gratitude is given to Dick Bennett (whose grandfather is featured throughout this book) for his early support, photographs, and oral history, without which I would have been left digging deeply for answers. The residents of Neihart, past and present, who shared time, stories, photographs, family manuscripts, and support for which I am so grateful include: Ruth Bach, Carol Zehntner, George Willett, Tony Juntunen, and any I have forgotten, thank you, too. I thank my cartographer, Michael Bugenstein of High Plains Mapping (www.highplainsmapping.com) in Glendive, Montana, for an outstanding job. Thanks to L.B. Stark, former Neihart mayor, who helped the Montana Historical Society with photograph identifications.

My Arcadia publisher editor, Jeff Ruetsche, put up with more issues blocking our way than most editors face in a lifetime. My wife, Dana Smith, gave unflagging support, rereading everything and editing late-night mistakes. And our daughter, Eryn, gently chided me through the years of low or no activity with, "So, how's the book coming along, Dad?"

I am also grateful to my dog Zoë, the incredible creature who would rather play stick or go snowshoeing for hours on end, but who understands that, when we go into the study and sit down at the computer, she is to take a (long) nap and patiently wait for the clicking and exhortations to end before she again is the center of my universe (at least as far as she knows).

The images in this volume appear courtesy of the History Museum for Cascade County and Great Falls, Montana (THM), Museum of the Rockies of Bozeman, Montana (MOR), Montana Historical Society, Helena (MHS), and the Minnesota Historical Society, Saint Paul (MnHS). Those from private collections are noted by the name of their contributors.

In the end, any errors are mine.

Introduction

Neihart's story has not been as well documented as those of other lesser-known mining districts in Montana. There are not a lot of diaries and letters preserved and waiting to be discovered in a dusty box in a museum or local historical society.

Throughout Neihart's 131 years, documentation and preservation of its history was not neatly packaged in a vault deep in the bowels of town hall. Since there is no evidence of an official town hall building, the best a researcher can do is mine the archives, dig through local and far-flung historical society records, and interview relatives of pioneering families to flesh out the history of Neihart's mines, residents, and growth, as well as its decline.

As readers of Western mining history know, mining starts out with an individual (or a small group of individuals) prospecting with basic equipment and confronting some rather harsh landscape that may hold something of great value for them. Ore-lode mining (hardrock mining) becomes an extremely complex operation, usually involving partners with capital, lawyers, assayers, and additional resources in order to reach some degree of success. Along the way, the complexity and expense of the process become even more critical factors. All this combined makes a fascinating story for those interested in Neihart, Montana, and mining.

Neihart's story is deserving of a photographic history book since pictures are able to speak of a past that mere words cannot make clear. This collection of images will share those early days of Neihart as it rose from a dusty and isolated mining camp of tents, log structures, and lean-tos, to become a village and, eventually, an incorporated town, filled with merchants, families, churches, schools, banks, saloons, cribs of prostitution, druggists, cafés, hotels, attorneys, restaurants, undertakers, boardinghouses, and a railroad.

If the land now called Montana or the mountain range now known as the Little Belt Mountains were to tell the story of how they came to be, it would be a dramatic tale. Continents collided. Volcanoes erupted. Glaciers marched across the land and retreated, and glacial lakes formed and drained time and again. Seas filled, overtopped their banks, and filled again. Winds blew dirt around, water carried dirt away, and earthquakes rearranged the dirt.

The Little Belt Mountains, just like all other ranges in the area, were formed millions of years ago, when pockets of magma pushed up (intruded) through the crust of the earth. Specifically, the Little Belt Mountains were bowed by numerous blister-like pockets of magma (molten rock) that pushed up or arched the sedimentary beds (layers) above them, forming domes. These domes and mountains were then incised by igneous rock, formed around 50 million years ago. The laccolithic dome underneath Neihart, its ore-lode mining area, begins at Belt Creek on the west, running on a northeast trend to the Dry Fork of Belt Creek and east to Big Baldy. It is geologically known as the Pioneer Ridge Uplift, running about 80 miles north to south and 30 miles west to east.

The intrusive host rocks were Precambrian (the oldest rocks on earth, dating back 3.8 billion years) and covered by Paleozoic (570 million to 245 million years ago) and Mesozoic (245 million to 65 million years ago) sediments with a thick series of lavas, probably of late Cretaceous (65 million to 50 million years ago) age. The structure of the host rocks is complex and beyond the geologic skills of an amateur to describe adequately.

The mineralization in the Neihart mining district took place during the early Tertiary period, about 65 million years ago. Faulting also occurred, and later came extensive erosion via wind, rain, and rock falls. The last volcanic igneous activity took place in the mid-Tertiary period (approximately 2 million years ago).

Long after these geologic processes, but long before there was a town or a mining camp, there were mountains covered in pines, gulches, and wildlife. For centuries, Native Americans passed through the area on their way to sacred grounds near the spiritual healing waters in present-day White Sulphur Springs, Montana. While we do not know that they hunted or fished in the many streams, we do know that the Blackfeet considered this their area long before Lewis and Clark viewed the Little Belt Mountains during their portage around the great falls more than 200 years ago.

Neihart was a product of the mining camp Barker to the north, in the Little Belt Mountains, and of the town of Fort Benton on the Missouri River, where the enterprising citizens were financially interested and tied to the prospects of Barker-Hughesville. Around July 1, 1881, James LeRoy Neihart, John C. O'Brien, and Richard Harley came into the gulch from Barker by way of Oti Park and Oti Park Creek, up, over, and down, to what later became known as Carpenter Creek and down towards Belt Creek. From that confluence, they spent two weeks bushwhacking to reach a promising outcropping that lead to the discovery of the Queen of the Hills Mine and other claims.

On April 7, 1882, with most of the prospectors in attendance, the tent/shack community held a meeting to decide on a name for the fledgling camp. James C. O'Brien was chairman, and D.B. Mackintosh (who would eventually become Neihart's first postmaster) was elected secretary. Mackintosh moved that the community should be called Farragut in honor of the famous US Navy admiral, but the motion was not seconded and was, therefore, not considered. Then, Charles Hamilton suggested the camp be known as Neihart. The motion was unanimously approved.

Neihart was once the hub of activity for other mining camps in the Little Belt Mountains. Following the 1881 discovery of silver, mining activity over the next 50 years fluctuated wildly, depending on the price of the metal. Due to the remoteness of the area, transportation of ore was problematic. A smelter was built near Neihart at the site of the Mountain Chief Mine, located northeast of Neihart, in the Carpenter Creek drainage. The smelter operated briefly from 1886 until 1887. A boon to the area was the 1891 completion of the Belt Mountain branch of the Montana Central Railroad, a subsidiary of the Great Northern Railroad, which connected Neihart to Great Falls and the Anaconda smelter so critically needed in the extraction of silver from its ore body.

Since Neihart was the railroad terminus, a round table was constructed on which the engine was turned around for the return trip to Monarch and points west, ending at Great Falls.

American Express Freight Company was among the first and most successful express delivery businesses to rise during the rapid westward expansion of the United States. The company got its start in Albany, New York, in 1850 and expanded by signing contracts with the first railroads in the Midwest and using boats to move freight around the country, including the newest US state at the time: California. The Wells Fargo Group also got its start in California as a transcontinental freight service, becoming a financial services business by setting up the first money orders. The money order process allowed customers to send money across the country in a safe way.

Winter presented serious challenges. The railroad was wise enough to firmly attach snowplows to the front of its steam engines. On more than one occasion, there were reports of blocked tracks. If that were not enough to make things miserable, the temperatures reached in the dead of winter certainly were. Although today's climate is milder, Montana suffered some of the worst winters of all time in the 1880s and 1890s, as depicted in Charlie Russell's famous painting, *Waiting for a Chinook*. Minus 40 degrees was attention getting. The winter of 1886 to 1887 contributed to the loss of hundreds of thousands of animals, mostly cattle, from Montana to Texas. Following a drought-filled summer, blizzards began in November and continued through February. By spring, dead cattle were everywhere—piled against fence lines, stacked in the coulees, and clogging rivers. Severe financial hardship and bankruptcy befell even the most solvent ranchers, including a young Theodore Roosevelt.

The railway line operated until 1945. Silver was the major ore extracted from the region; however, zinc, galena, lead, and gold were also mined. Fascinating people of diverse ancestry, the early mining residents of Neihart are the focus of this work. They are seen in these fascinating photographs working their mining claims, marching through muddy streets, and picnicking alongside Belt Creek. These images reflect the lay of the camp/town as it first followed the veins of ore, the flow of Belt Creek, and, eventually, the path of the railroad tracks. Acquired from the collection of the History Museum of Cascade County and Great Falls, from descendants of pioneers, and from the photograph archives of the Montana Historical Society and several other archives, these photographs help illustrate the progression of a mining camp into a town and an important producer of silver.

Author and amateur historian Ronald L. Lansverk (Ron) is a 30-year resident (some years part-time) of the Neihart community and has almost completed his 29-year quest to unearth the history of this once-booming silver-mining community of the late 19th century.

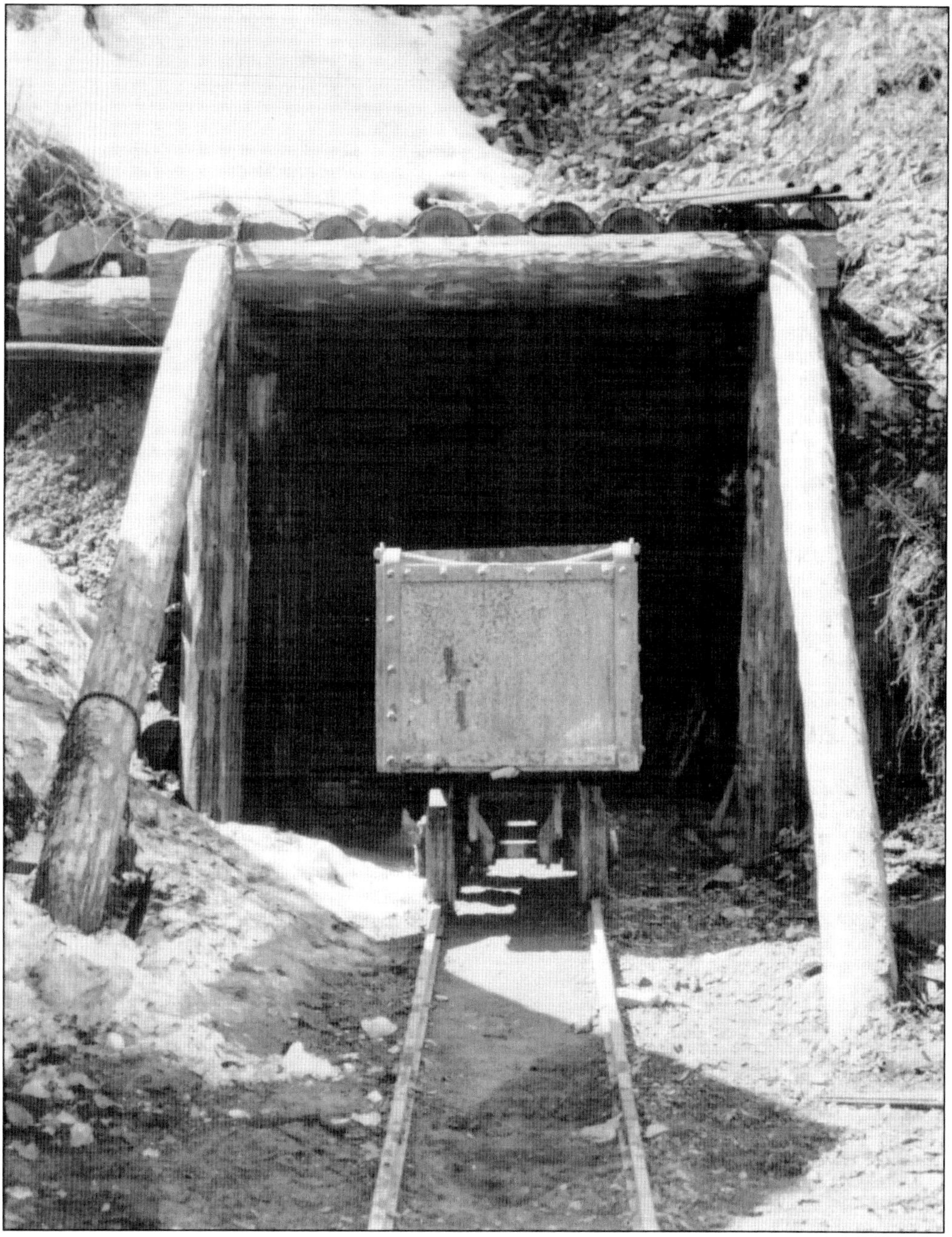

ORE CAR WITH MULE POWER. The ore car from an unidentified silver mine in the Neihart area is a remarkable identifier for an industry. It is doubtful that any person in a far west state with a mining history would not be able to immediately identify that piece of equipment. This is, after all, a reminder of the importance of mining to Montana's historic development and economy. The ore car serves as a reminder that the route from silver mine to silver dollar was long and dangerous at every step, but it is also a reminder that that is what is loved about Montana—the rugged individuals it took to mine these hills and rocks in order to make a better place for the future. (MHS.)

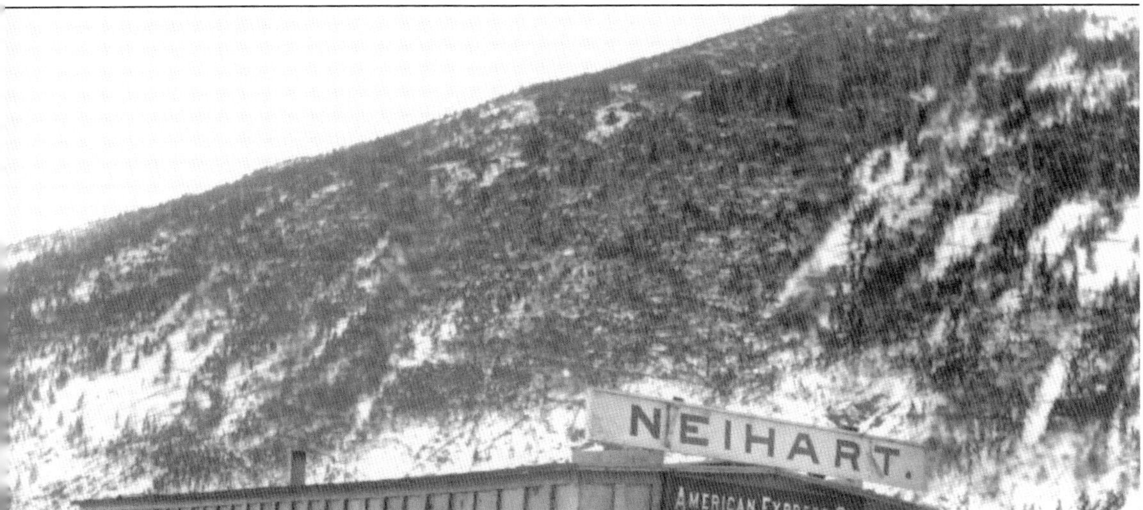

NEIHART'S FIRST FREIGHT HOUSE BY AMERICAN EXPRESS. This photograph was taken at the American Express freight house depot in Neihart c. 1893. The three unknown gentlemen are perhaps officials of American Express from Great Falls, along with officials from the Neihart depot or Montana Central Railroad. The depot was located at the terminus of the rail line in an area near the round table that was located just north of the current Neihart-Monarch Senior Citizens Center, which was originally the second Neihart School. The area now is mostly marshy, with Belt Creek creating an area of wetlands. (MHS.)

One

Discovery

A prospector lives in hopes, and dies in some old cabin.

—Hutch Stevens, in *The Restless Longing: A Prospector's Story*

Prospectors from the Barker/Hughesville mining district are credited with making the first discoveries in the Neihart district in 1881, after the excitement from the initial discoveries in the Barker district had come to an end. The first claim in the Neihart district was made in July 1881 at the Queen of the Hills Mine. This mine was a rich early producer of silver ore. Activity continued in the Neihart mining district in 1883 and 1884 with the development of additional mines, including the Galt, the Mountain Chief, and the Silver Belt.

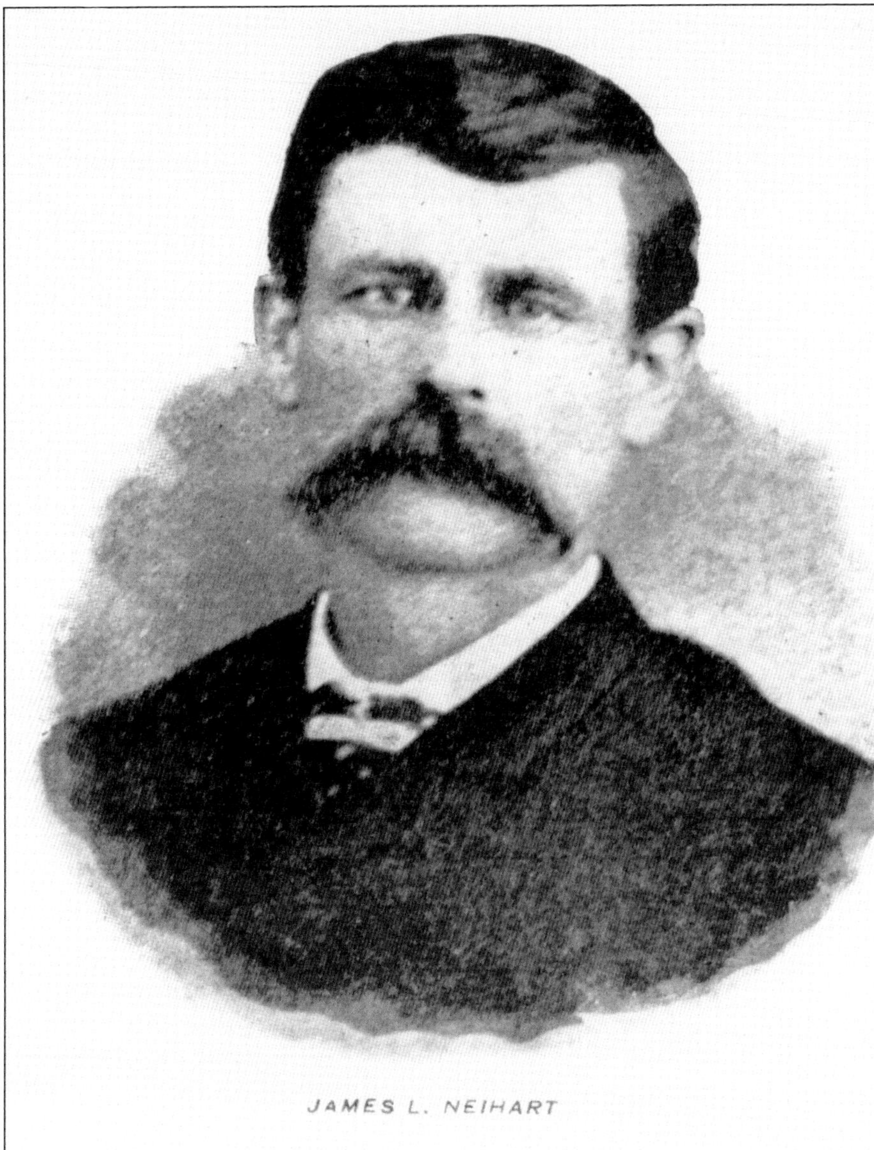

JAMES L. NEIHART. James L. Neihart accomplished a lot during his lifetime, but not much of it was recorded after he struck silver, married his wife, Sarah, and was elected mayor of Neihart. He spent the rest of his years making mining investments and enjoying his life in the Little Belt Mountains of Montana. Neihart died on March 4, 1894. At the time of his death, he had been preparing new mining ventures near Libby, Montana. James was born at Fort Byron, Ohio, on September 1, 1851, as one of seven children. In 1876, at age 25, he moved Bozeman. An educated man, he taught one term of school before prospecting the Virginia City and Alder Gulch area for two years. In 1879, he moved to the area of Fort Benton and the Highwood Mountains, making a squatter's claim until 1881. He traded his holdings for a horse and some cash and struck out for Barker, then a booming camp. Neihart and two others, John O'Brien and Richard Harley, crossed over to the yet unnamed Carpenter Creek (by way of the Oti Park trail) hoping to see shinier rocks than they had in Barker. Apparently they did, because around July 1, 1881, they struck silver and staked claim to the Queen of the Hills. (THM.)

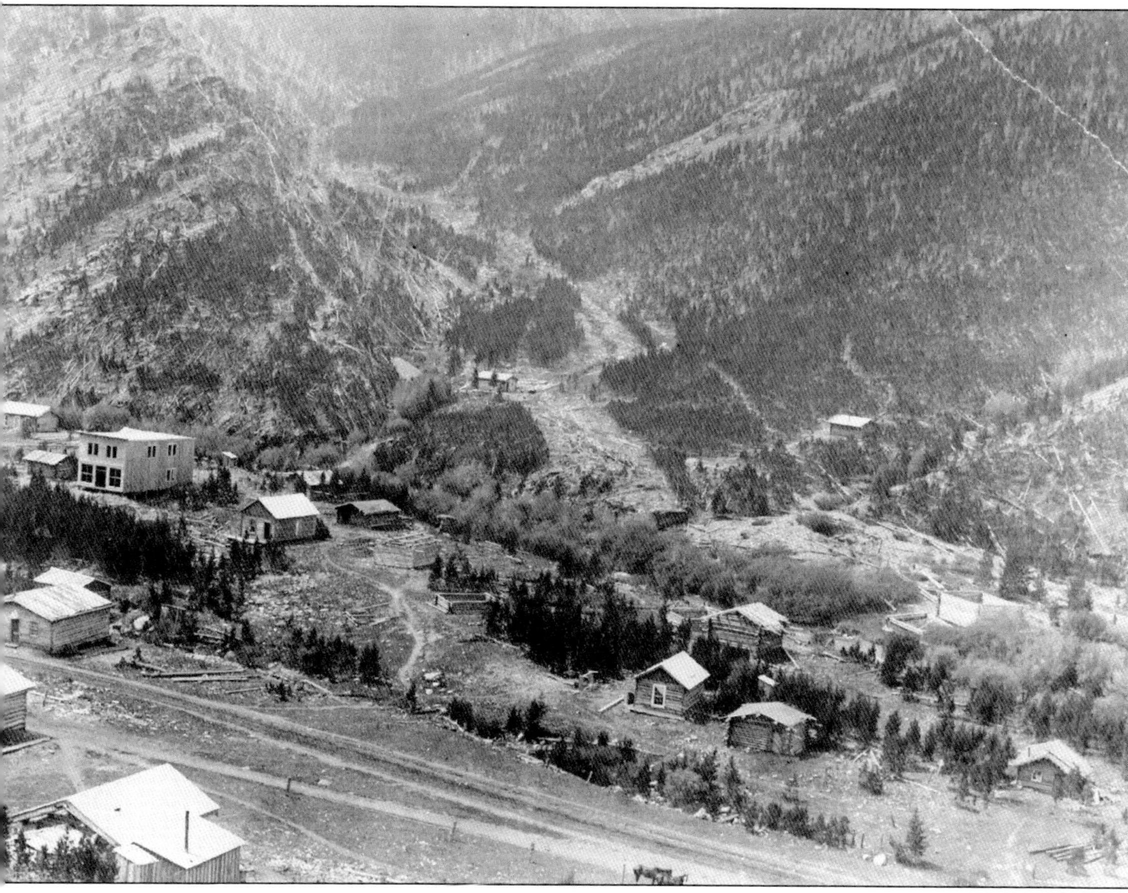

EARLY DAYS OF NEIHART. Mining camps, if anything, held amazing recuperative powers. With confirmation around 1889 or 1890 that the railroad was coming to Neihart, the camp sprang back to life. People flocked in, mines changed hands, and vacant lots suddenly became valuable. Hammers and saws provided background noise; Neihart was experiencing an actual building boom and enjoying every moment. All that good news—the building frenzy and promise of better times—provided the impetus to petition Meagher County to permit Neihart to incorporate and become a town. That happened towards the end of November 1891, and on December 4, 1891, James L. Neihart was elected as the city's first mayor. On March 18, 1892, the *Great Falls Tribune* detailed a personal report by Great Falls Water Works manager E.W. King on Neihart's recent growth, which read in part, "there are 166 residence houses in the town, 31 of which are two stories and contain 4 rooms or over; 135 are of one story and contain less than 4 rooms. The town has 35 stores, 22 saloons, 4 hotels, 11 restaurants, 6 laundries and 4 barber shops." That report reflected a huge increase from two years previous; less than a year later, the bottom fell out of the silver market for good. (MHS.)

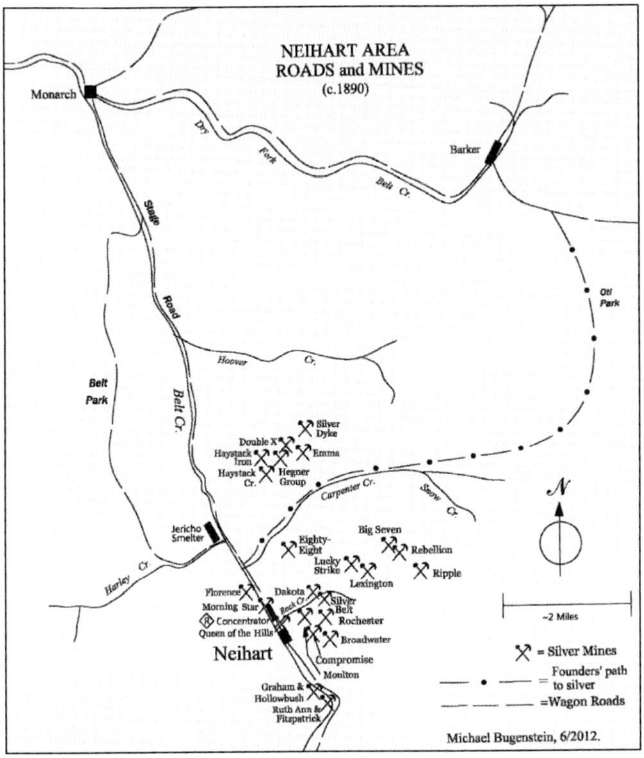

Area Foot and Wagon Trails. Many have long wondered how those first prospectors reached Canyon City (present-day Neihart). And how did Neihart, O'Brien, and Harley get from Barker to the Queen of the Hills lode? The map shows the founders' path to silver was by way of Oti Park Creek and Oti Park. This led them to what is now known as Carpenter Creek, which they followed down to its confluence with Belt Creek. It is said that, from that point, it took the men two weeks of bushwhacking to reach the clearings near present-day Neihart. (Author's collection.)

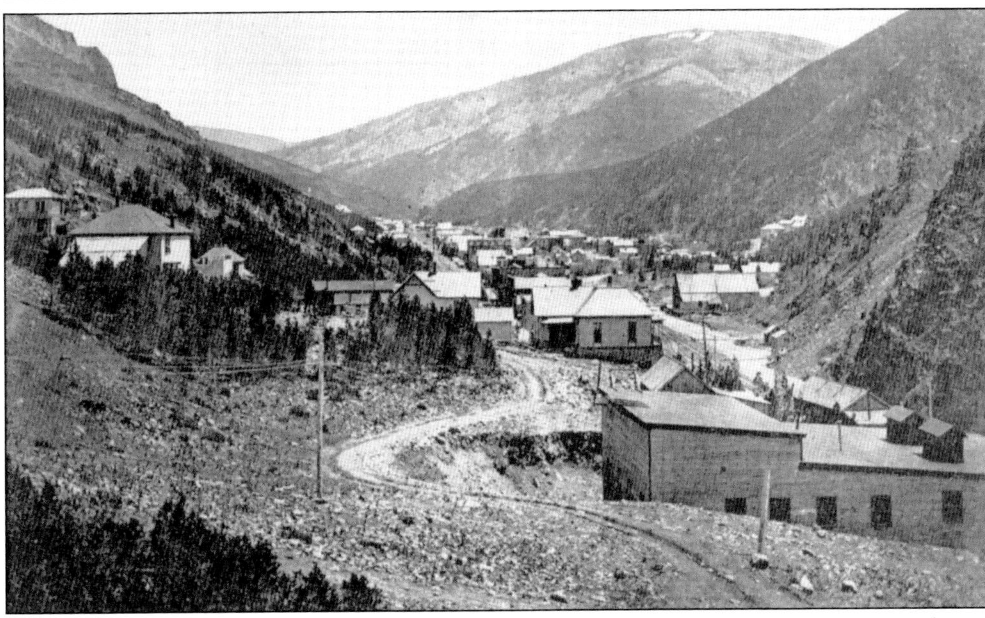

Original Entrance into Neihart. This was taken before the railroad arrived, around 1889 or 1890. Entering Neihart was not always a simple or straight proposition. The wagon road first passed the Florence Mine and then meandered around, above, and below the Queen of the Hills, which sat prominently in the way. The curve went up Rock Creek and then back towards Belt Creek before winding back up and, finally, down to the flatter portion of the gulch and Main Street. (MHS.)

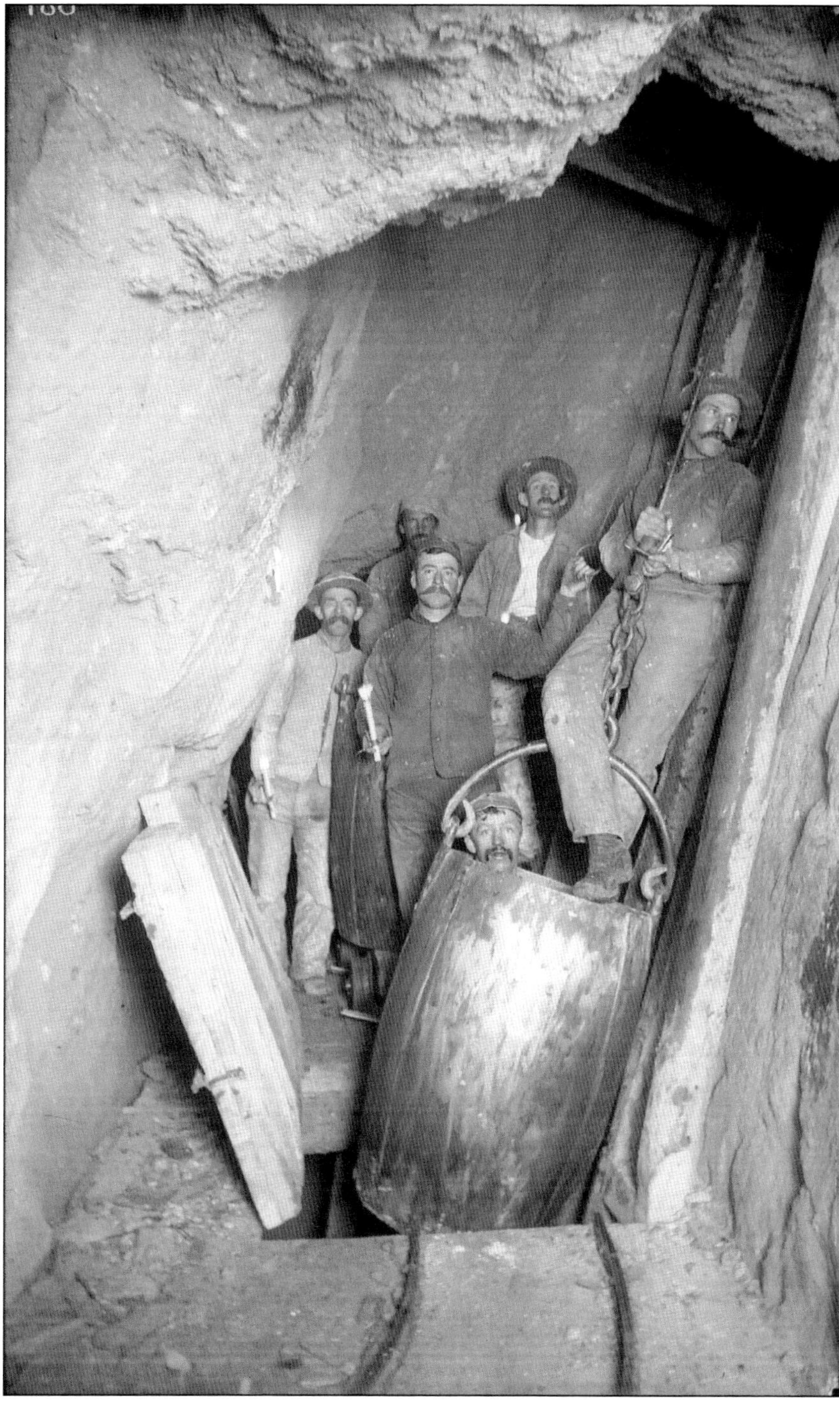

DESCENDING OR RISING TO A WORK LEVEL. This group of unidentified hardrock miners gathers to either descend or ascend to another level (each mining level was 100 feet below the previous one), either beginning or ending their shift in the mine. Some referred to this as "riding the bucket," whereas others used more colorful descriptions. The slant of the bucket, about 25 degrees, allowed the miners to follow the vein as they worked lower in the earth. (Colorado Historical Society.)

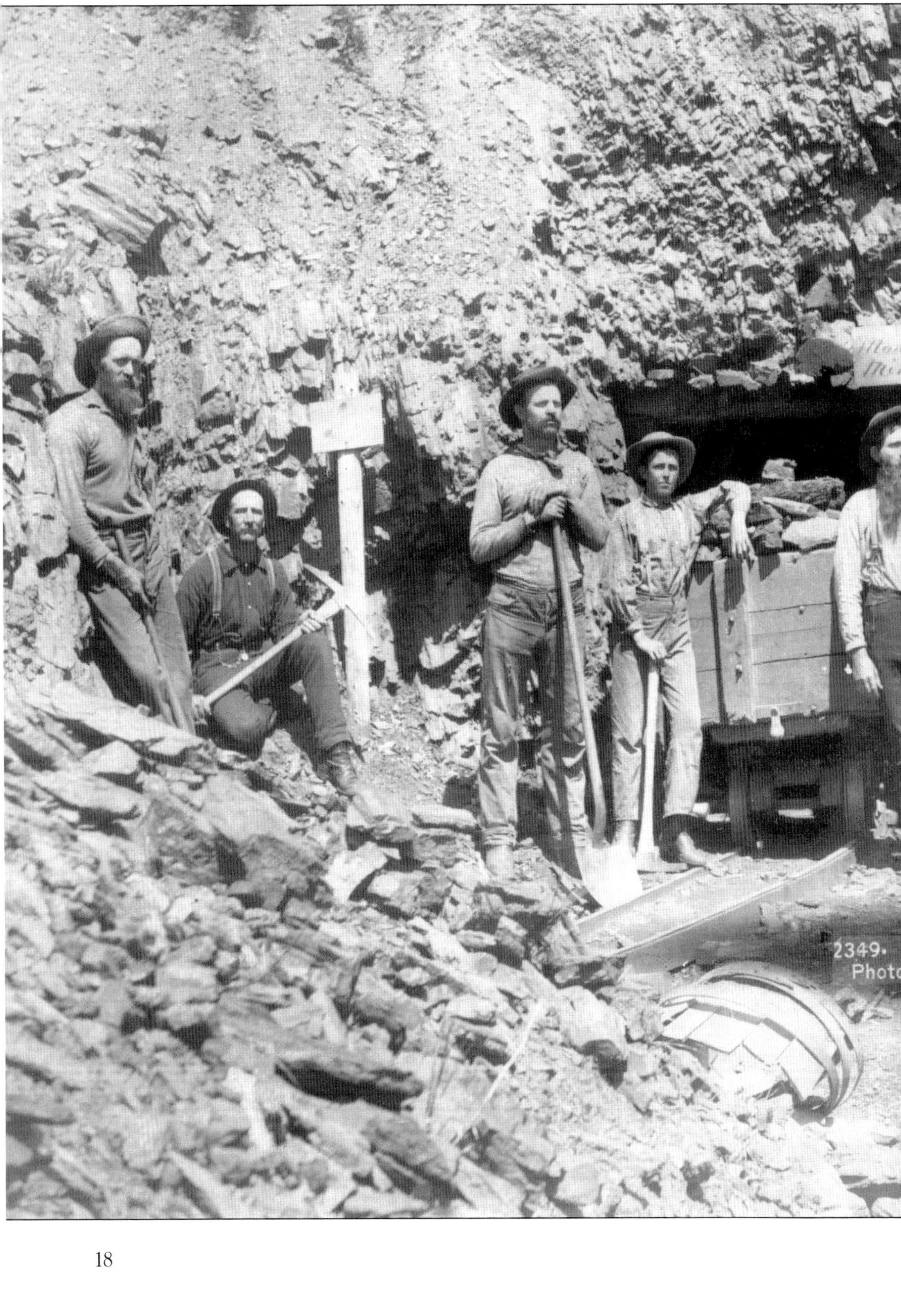
2349.

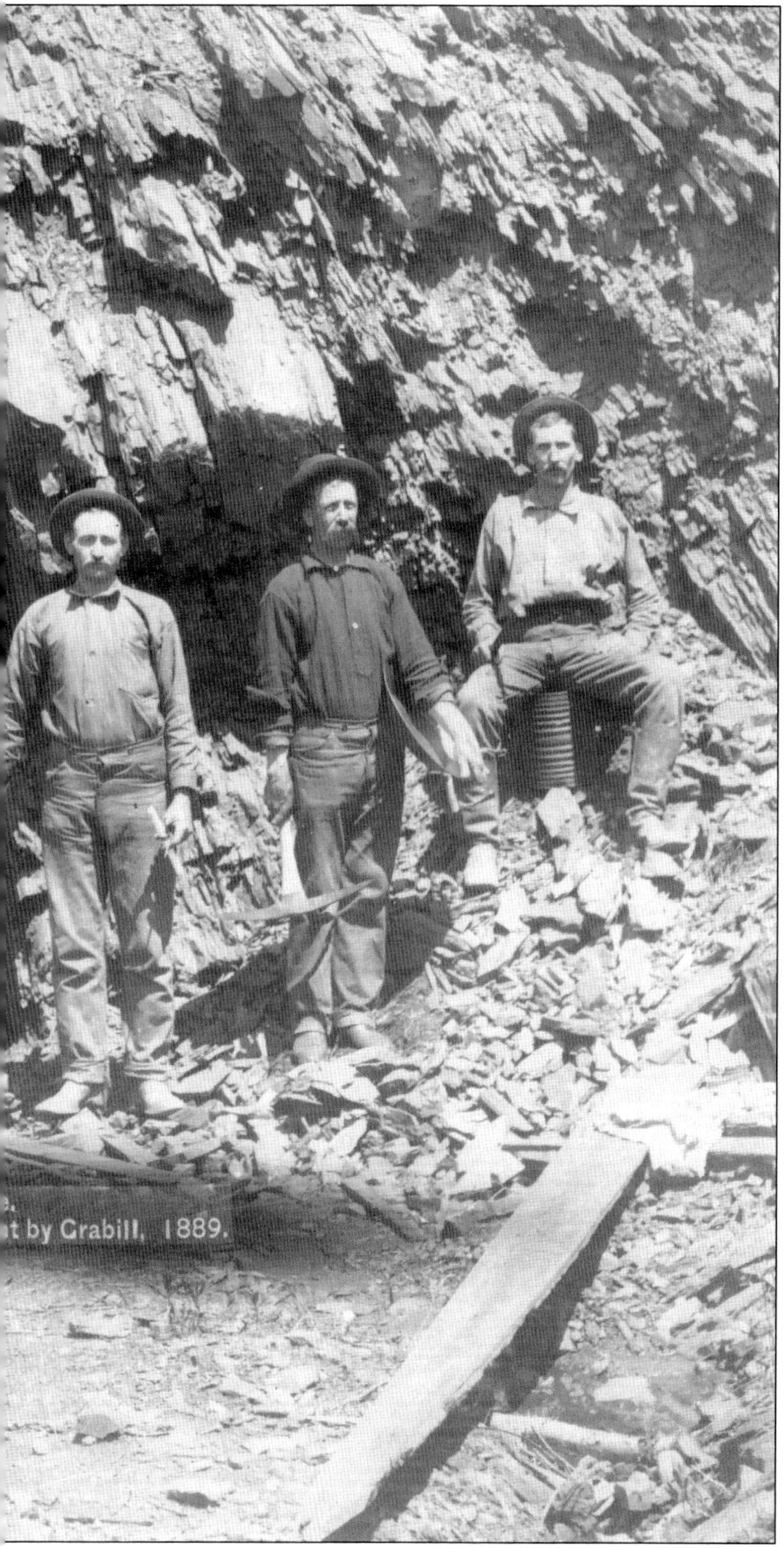

Early Work by Early Crews. This Montana mine is photographed in 1889 by John C.H. Grabill. Between 1887 and 1892, Grabill sent 188 photographs to the Library of Congress for copyright protection. Grabill is known as a Western photographer, who documented many aspects of frontier life, including hunting, mining, town landscapes, and white settlers' relationships with Native Americans. This photograph shows the rugged terrain and conditions the early prospectors faced. (John C.H. Grabill Collection, LOC.)

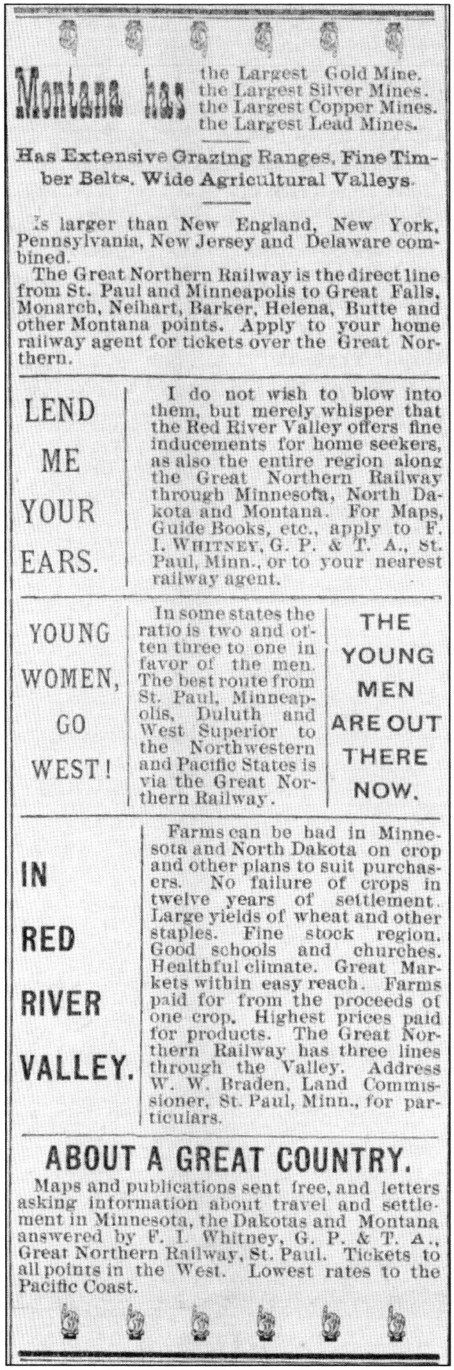

Montana has the Largest Gold Mine.
the Largest Silver Mines.
the Largest Copper Mines.
the Largest Lead Mines.

Has Extensive Grazing Ranges, Fine Timber Belts, Wide Agricultural Valleys.

Is larger than New England, New York, Pennsylvania, New Jersey and Delaware combined.

The Great Northern Railway is the direct line from St. Paul and Minneapolis to Great Falls, Monarch, Neihart, Barker, Helena, Butte and other Montana points. Apply to your home railway agent for tickets over the Great Northern.

LEND ME YOUR EARS.

I do not wish to blow into them, but merely whisper that the Red River Valley offers fine inducements for home seekers, as also the entire region along the Great Northern Railway through Minnesota, North Dakota and Montana. For Maps, Guide Books, etc., apply to F. I. Whitney, G. P. & T. A., St. Paul, Minn., or to your nearest railway agent.

YOUNG WOMEN, GO WEST!

In some states the ratio is two and often three to one in favor of the men. The best route from St. Paul, Minneapolis, Duluth and West Superior to the Northwestern and Pacific States is via the Great Northern Railway.

THE YOUNG MEN ARE OUT THERE NOW.

IN RED RIVER VALLEY.

Farms can be had in Minnesota and North Dakota on crop and other plans to suit purchasers. No failure of crops in twelve years of settlement. Large yields of wheat and other staples. Fine stock region. Good schools and churches. Healthful climate. Great Markets within easy reach. Farms paid for from the proceeds of one crop. Highest prices paid for products. The Great Northern Railway has three lines through the Valley. Address W. W. Braden, Land Commissioner, St. Paul, Minn., for particulars.

ABOUT A GREAT COUNTRY.

Maps and publications sent free, and letters asking information about travel and settlement in Minnesota, the Dakotas and Montana answered by F. I. Whitney, G. P. & T. A., Great Northern Railway, St. Paul. Tickets to all points in the West. Lowest rates to the Pacific Coast.

Railroad Advertising Draws Settlers to Area. "You can lay track to the Garden of Eden, but what good is it if the only inhabitants are Adam and Eve?" asked one railroad executive in the mid-1880s. With that simplistic business reasoning, the railroads began promoting the areas they served. After all, what purpose do the railroads serve, if not to move freight and merchandise into the territory and export farming, ranching, and mineral production out of it? Without customers, the Great Northern Railroad had no business in Montana. (LOC.)

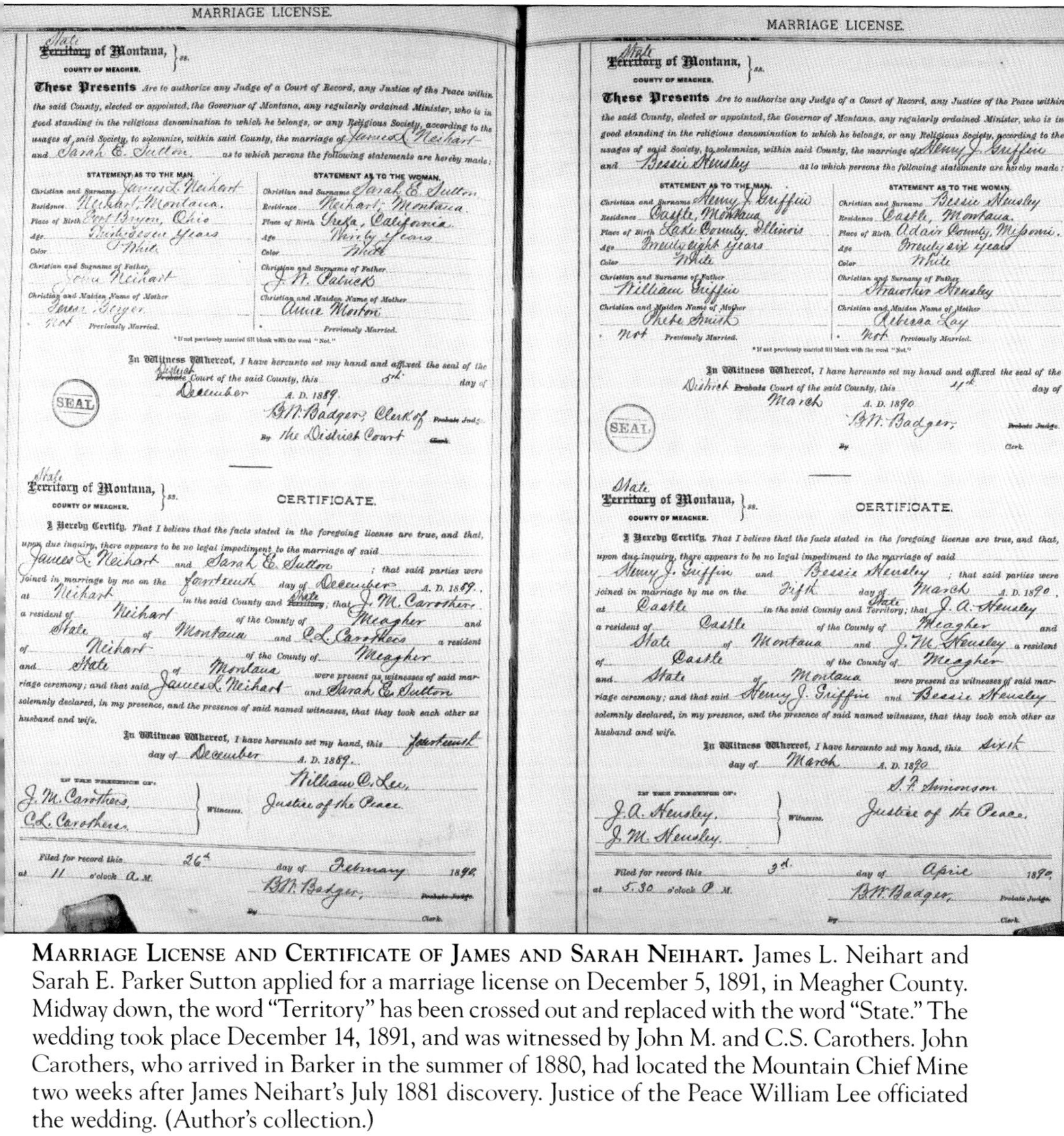
MARRIAGE LICENSE.

~~Territory~~ State of Montana, } ss.
COUNTY OF MEAGHER.

These Presents Are to authorize any Judge of a Court of Record, any Justice of the Peace within the said County, elected or appointed, the Governor of Montana, any regularly ordained Minister, who is in good standing in the religious denomination to which he belongs, or any Religious Society, according to the usages of said Society, to solemnize, within said County, the marriage of James L. Neihart and Sarah E. Sutton as to which persons the following statements are hereby made:

STATEMENT AS TO THE MAN.	STATEMENT AS TO THE WOMAN.
Christian and Surname James L. Neihart	Christian and Surname Sarah E. Sutton
Residence Neihart, Montana.	Residence Neihart, Montana
Place of Birth Port Bryon, Ohio	Place of Birth Yreka, California
Age Thirty-Seven years	Age Thirty years
Color White	Color White
Christian and Surname of Father John Neihart	Christian and Surname of Father J. N. Patrick
Christian and Maiden Name of Mother Susan Geiger	Christian and Maiden Name of Mother Annie Morton
Not Previously Married.	Previously Married.

*If not previously married fill blank with the word "Not."

In Witness Whereof, I have hereunto set my hand and affixed the seal of the District ~~Probate~~ Court of the said County, this 5th day of December A. D. 1889.

SEAL

B.N. Badger, Clerk of ~~Probate Judge.~~
By the District Court ~~Clerk.~~

~~Territory~~ State of Montana, } ss.
COUNTY OF MEAGHER.

CERTIFICATE.

I Hereby Certify, That I believe that the facts stated in the foregoing license are true, and that, upon due inquiry, there appears to be no legal impediment to the marriage of said James L. Neihart and Sarah E. Sutton; that said parties were joined in marriage by me on the fourteenth day of December A. D. 1889. at Neihart in the said County and ~~Territory~~ State; that J. M. Carothers a resident of Neihart of the County of Meagher and State of Montana and C.L. Carothers a resident of Neihart of the County of Meagher and State of Montana were present as witnesses of said marriage ceremony; and that said James L. Neihart and Sarah E. Sutton solemnly declared, in my presence, and the presence of said named witnesses, that they took each other as husband and wife.

In Witness Whereof, I have hereunto set my hand, this fourteenth day of December A. D. 1889.

IN THE PRESENCE OF:
J. M. Carothers,
C.L. Carothers. } Witnesses.

William C. Lee,
Justice of the Peace

Filed for record this 26th day of February 1890
at 11 o'clock a. M.
B.N. Badger, ~~Probate Judge.~~
By ~~Clerk.~~

MARRIAGE LICENSE.

~~Territory~~ State of Montana, } ss.
COUNTY OF MEAGHER.

These Presents Are to authorize any Judge of a Court of Record, any Justice of the Peace within the said County, elected or appointed, the Governor of Montana, any regularly ordained Minister, who is in good standing in the religious denomination to which he belongs, or any Religious Society, according to the usages of said Society, to solemnize, within said County, the marriage of Henry J. Griffin and Bessie Hensley as to which persons the following statements are hereby made:

STATEMENT AS TO THE MAN.	STATEMENT AS TO THE WOMAN.
Christian and Surname Henry J. Griffin	Christian and Surname Bessie Hensley
Residence Castle, Montana	Residence Castle, Montana.
Place of Birth Lake County, Illinois	Place of Birth Adair County, Missouri.
Age Twenty eight years	Age Twenty six years
Color White	Color White
Christian and Surname of Father William Griffin	Christian and Surname of Father Strawther Hensley
Christian and Maiden Name of Mother Phebe Smith	Christian and Maiden Name of Mother Rebecca Lay
Not Previously Married.	Not Previously Married.

*If not previously married fill blank with the word "Not."

In Witness Whereof, I have hereunto set my hand and affixed the seal of the District ~~Probate~~ Court of the said County, this 4th day of March A. D. 1890.

SEAL

B.N. Badger, ~~Probate Judge.~~
By Clerk.

State ~~Territory~~ of Montana, } ss.
COUNTY OF MEAGHER.

CERTIFICATE.

I Hereby Certify, That I believe that the facts stated in the foregoing license are true, and that, upon due inquiry, there appears to be no legal impediment to the marriage of said Henry J. Griffin and Bessie Hensley; that said parties were joined in marriage by me on the Fifth day of March A. D. 1890. at Castle in the said County and State Territory; that J. A. Hensley a resident of Castle of the County of Meagher and State of Montana and J. M. Hensley a resident of Castle of the County of Meagher and State of Montana were present as witnesses of said marriage ceremony; and that said Henry J. Griffin and Bessie Hensley solemnly declared, in my presence, and the presence of said named witnesses, that they took each other as husband and wife.

In Witness Whereof, I have hereunto set my hand, this sixth day of March A. D. 1890

IN THE PRESENCE OF:
J. A. Hensley,
J. M. Hensley. } Witnesses.

S. F. Simonson
Justice of the Peace.

Filed for record this 3d. day of April 1890,
at 5.30 o'clock P. M.
B.N. Badger, Probate Judge.
By Clerk.

MARRIAGE LICENSE AND CERTIFICATE OF JAMES AND SARAH NEIHART. James L. Neihart and Sarah E. Parker Sutton applied for a marriage license on December 5, 1891, in Meagher County. Midway down, the word "Territory" has been crossed out and replaced with the word "State." The wedding took place December 14, 1891, and was witnessed by John M. and C.S. Carothers. John Carothers, who arrived in Barker in the summer of 1880, had located the Mountain Chief Mine two weeks after James Neihart's July 1881 discovery. Justice of the Peace William Lee officiated the wedding. (Author's collection.)

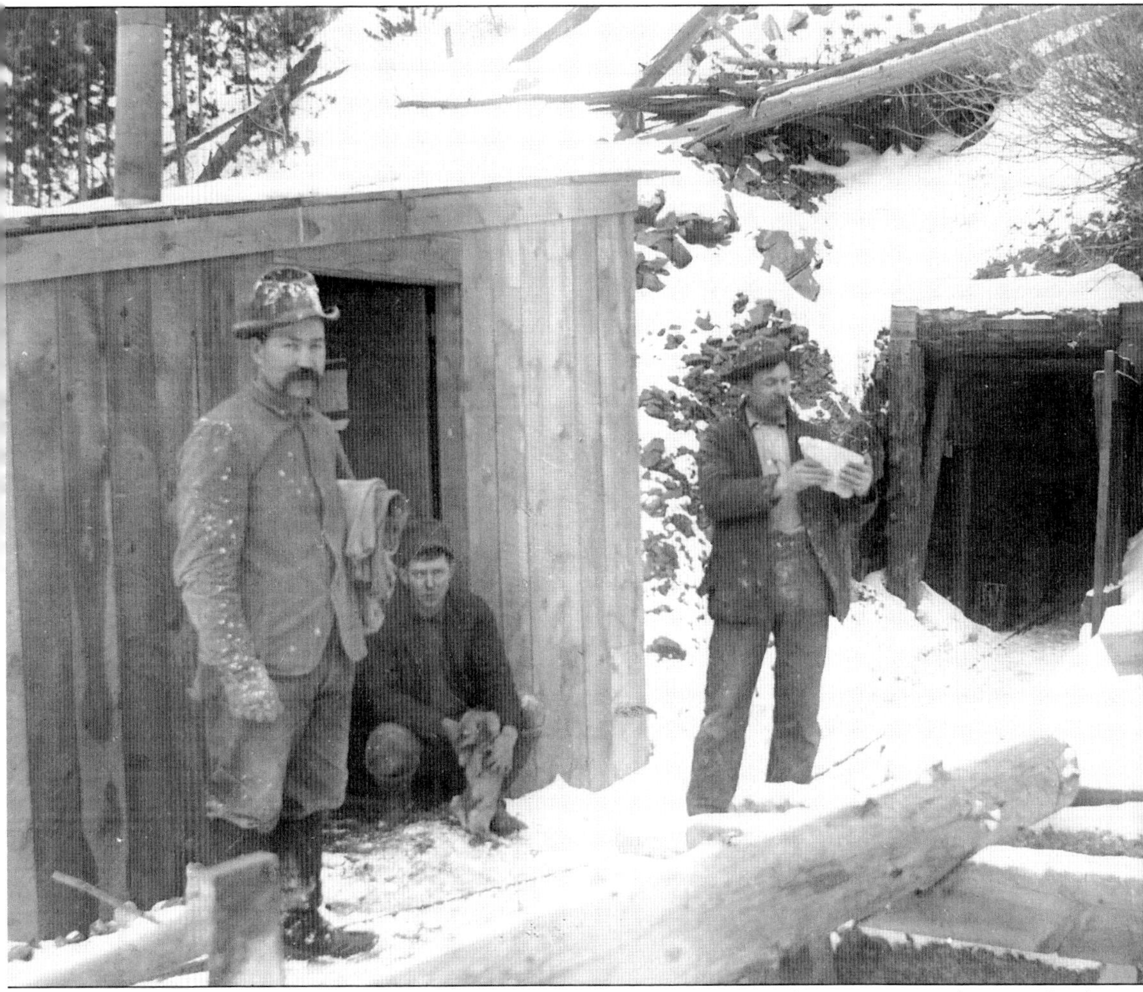

GETTING READY FOR THE JOB. Weather was a continual concern for miners, merchants, schools, parents, and railroads. The railroad attached snow plows to the front of its steam engines, and on more than one occasion, there were reports of blocked tracks. Temperatures dipped to minus 40 degrees. Montana suffered some of the worst winters of all time in the 1880s and 1890s. Charlie Russell's famous painting *Waiting for a Chinook* showed harsh winters in central Montana. The only population that prospered during those harsh winters was the wolves. These unidentified miners are pictured with their "choppers" (mittens) and boots. (MHS.)

Two

Mines, Miners, and Owners

The Only Chance a miner had of getting into heaven was to start a gold rush there.

—Richard E. Lingenfelter, *The Hardrock Miners*

Hardrock miners were a different breed from those who had roamed the hills, valleys, and creek and river bottoms of Alder Gulch, Last Chance Gulch, or Confederate Gulch. Their coveted gold or silver was encased in rocks, deep in the earth, requiring a large amount of digging, hauling, and breaking to be repeated day after day. But in the late 1880s and 1890s, the hardrock miner was looked upon as one to be emulated, having risen to the metaphorical major leagues of mining.

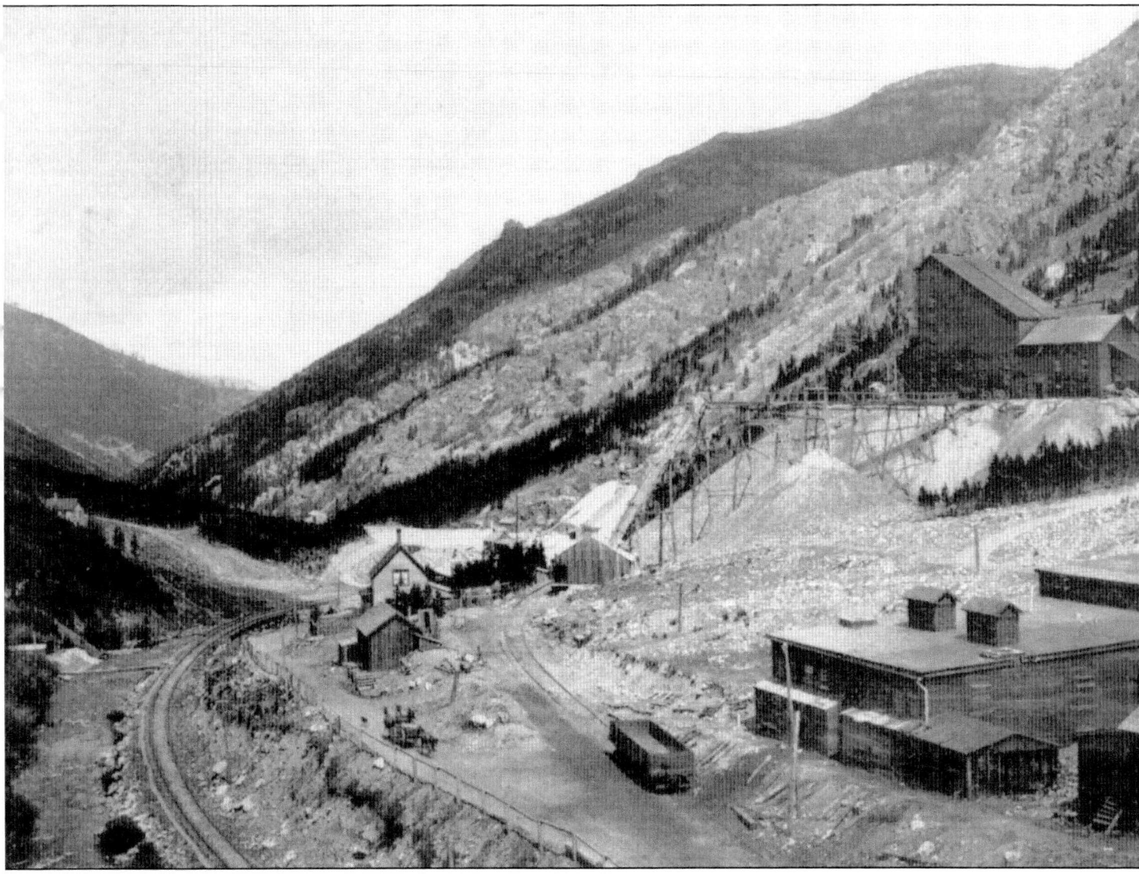

QUEEN OF THE HILLS OVERLOOKS NEIHART, C. 1895. As this photograph shows, the entrance into Neihart was not easy for man, horse, wagon, or a buggy with a family. The Queen, the first discovery in the gulch, sits as if guarding the canyon's deeply buried treasures from view. If one wanted entrance, one had to work mightily for access. Notice the lack of room between the Queen, the tracks, and Belt Creek. Looking carefully today, one can recognize that corner in Neihart. (THM.)

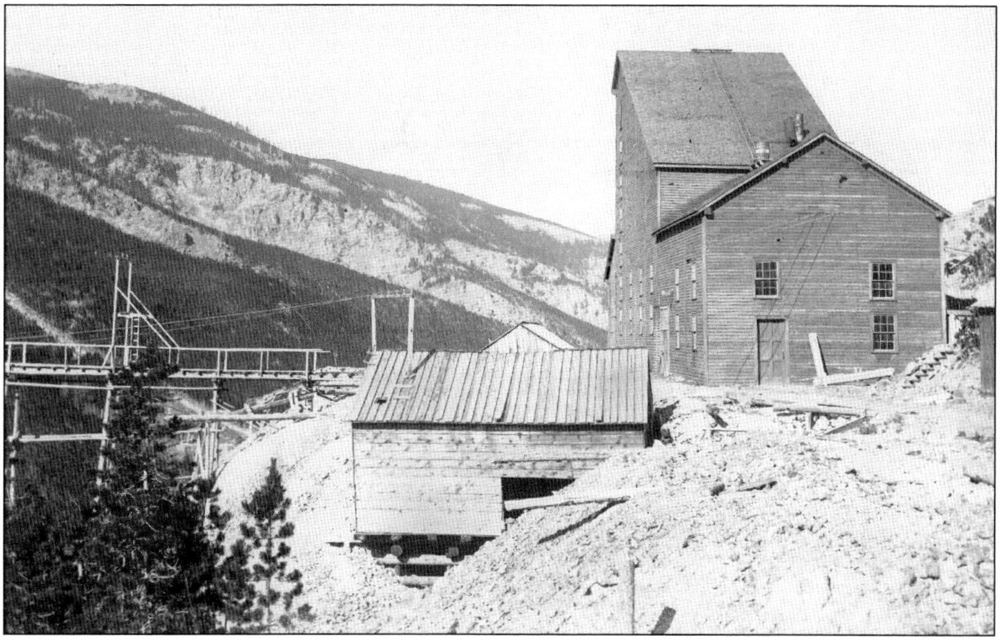

Queen of the Hills Mine Stands Ready For Electrical Power. On July 12, 1891, the *Great Falls Tribune* reported that the Queen of the Hills management and the Thomson-Houston Electric Light Company of Boston had agreed to a $50,000 contract, marking an epoch in Montana mining. "It was the beginning of electrical power operating mining equipment and powering the community," according to the *Tribune*. (MHS.)

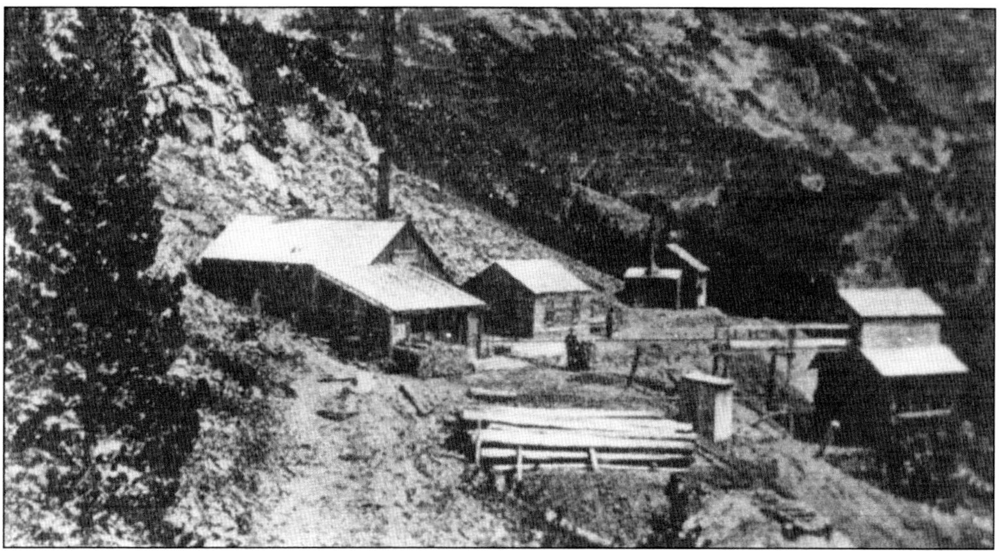

The Galt Works. Once called the "Mining Wonder" in camp, this mine was said to be working a continuous ore vein over 1,100 feet long. The state mining inspector's notes did not mention that wondrous vein, but he did write that this "property had produced a large quantity of high-grade ore during the years," indicating that it would continue to do so. Lying east of the O'Brien Mine, it was owned in 1895 by C.D. Ladd, Frank Marion, E.W. Toole, and Jonathan McAssey (who had added responsibilities as superintendent–general manager). McAssey and foreman Andy Nelson directed 25 men. (THM.)

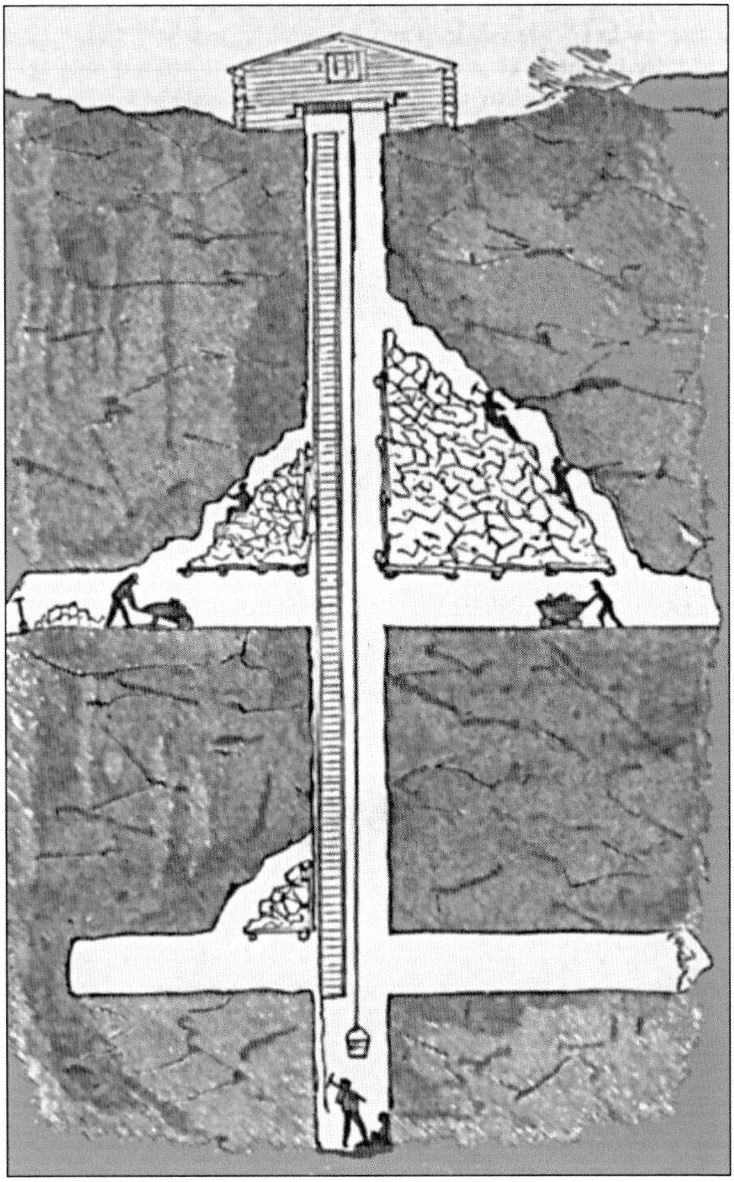

A Basic Hardrock Mine. Trying to visualize mining without a visual aid is like trying to learn the violin while appearing on stage. This simplified diagram shows a basic shaft, or winze, used to lower miners into the working area by means of a windlass, a device for hoisting. The image also depicts both a drift and a stope; the miners move the ore to the main shaft through the drift after mining the stope, where the vein is located. (*Mining and Scientific Press,* January 18, 1873.)

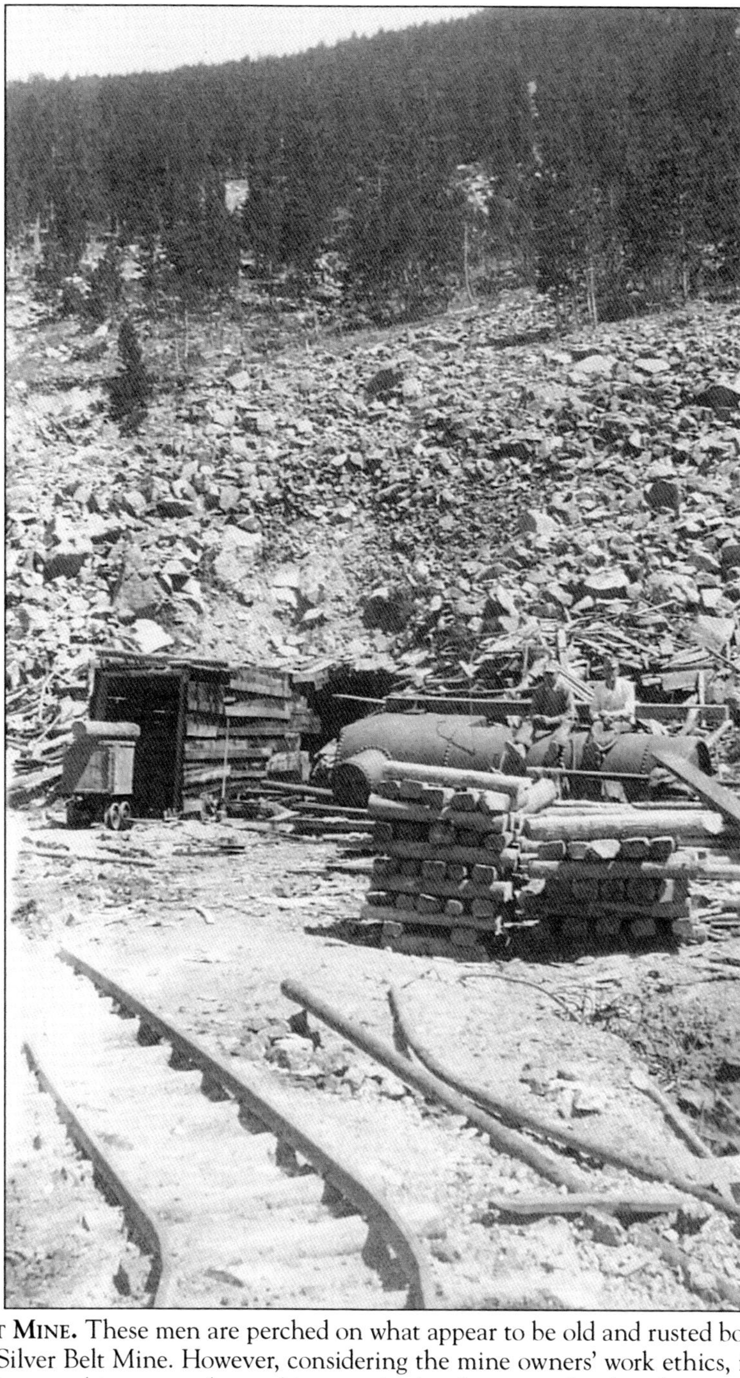

Silver Belt Mine. These men are perched on what appear to be old and rusted boilers near an adit to the Silver Belt Mine. However, considering the mine owners' work ethics, it is doubtful they are doing anything more than taking a quick break to pose for this photograph. James L. Neihart and Jonathan McAssey described the Silver Belt as having a 2,500-foot tunnel working a vein six to eight feet in width. They claimed to have found rich float (loose rock with ore lying around) on the ground and, in many places, pay ore at the grass roots (indicating profitable ore was just below the surface at grass root level.) The claim was a close neighbor to the Dacotah Mine and was 3,000 feet through Baldy from the Broadwater vein. (THM.)

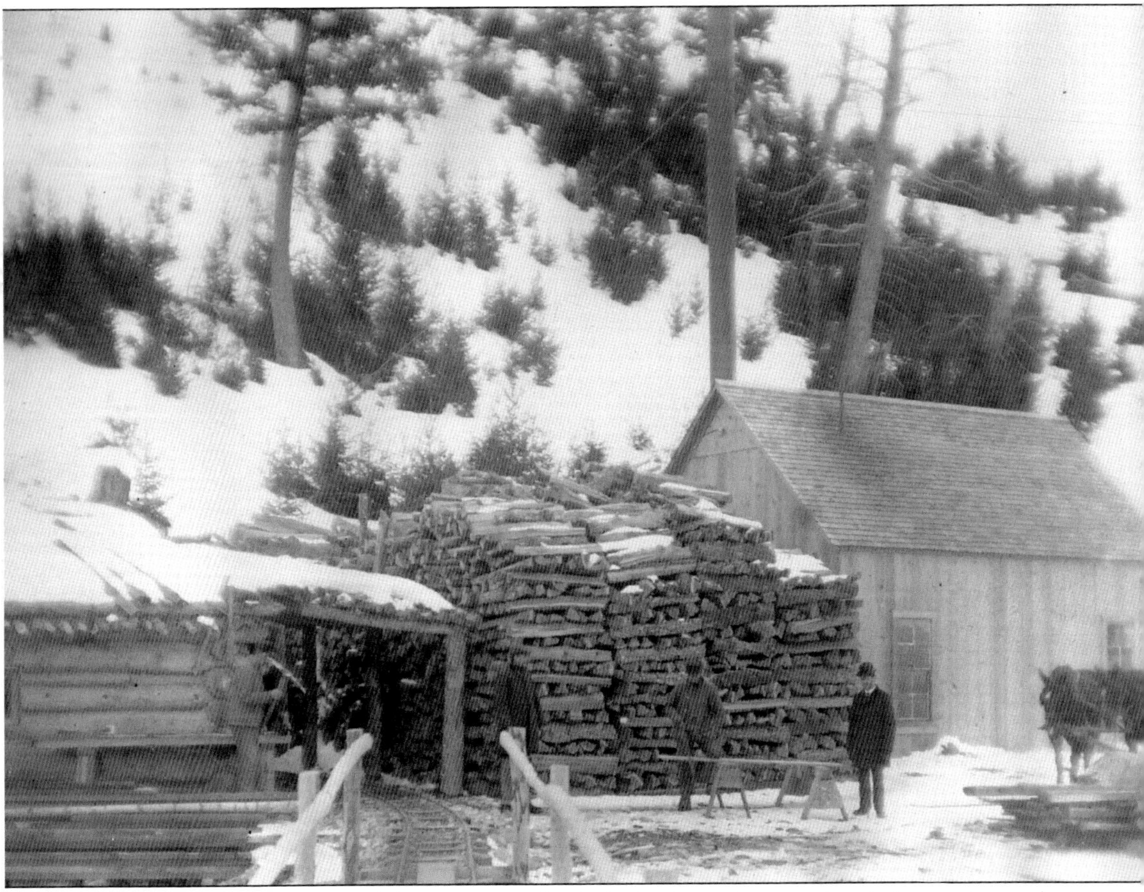

Barker Mine Sawmill. This photograph of the Barker mines in the Neihart mining district represents an everyday chore for most mine operators. Wood was used for many purposes in a mining operation, such as timbering in the mines, crosscut blocking, ties for rails, fuel for heat and steam, and reinforcement walls. One of the first concerns many prospectors had when encountering the Neihart area was the lack of timber. While there was timber all over the area, it was preferable to have the supply on, or as close as possible to, the property so that overhead costs and time spent away from breaking rock could be reduced. When viewing Neihart today, it can be hard to believe that there was not a lot of timber in the area. However, photographs from the 1880s and 1890s show that Neihart looked quite different then. The road south towards Showdown and King's Hill Pass was originally called Sawmill Road because that was the area where independent loggers found their most reliable supply of timber. That portion of Belt Creek was briefly known as "Sawmill Creek" until it became clear that it actually served as the headwaters of Belt Creek. (THM.)

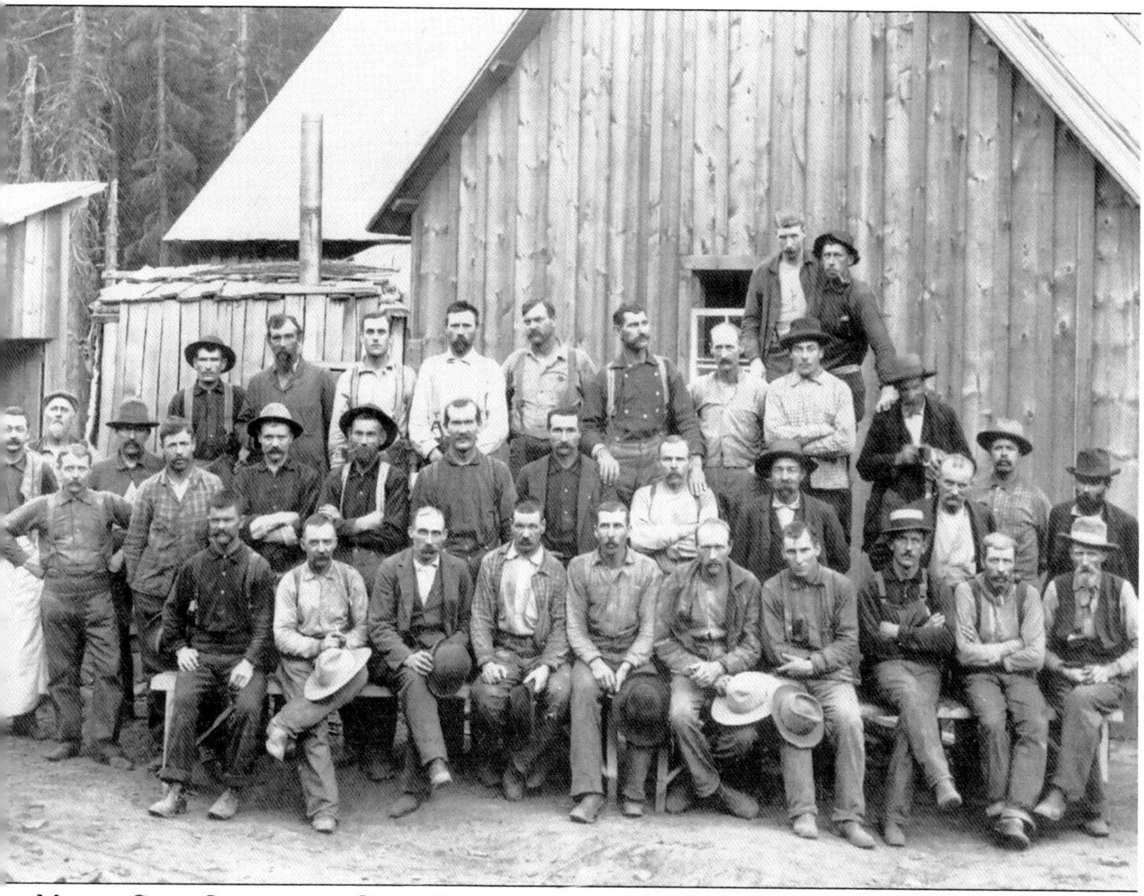

MINING CREW GATHERS FOR GROUP PHOTOGRAPH. Hordes of men, representing a variety of ethnicities, educational levels, and religions, headed west in the mid- to late 19th century. They all hoped to strike it rich; whether any of these unidentified individuals were successful in this endeavor is unknown. However, back then, photographic subjects had to remain motionless in order to not blur the final photograph. Because they could not be sure that they would be able to maintain the expression for the entire length of the exposure time, nobody smiled. (MHS.)

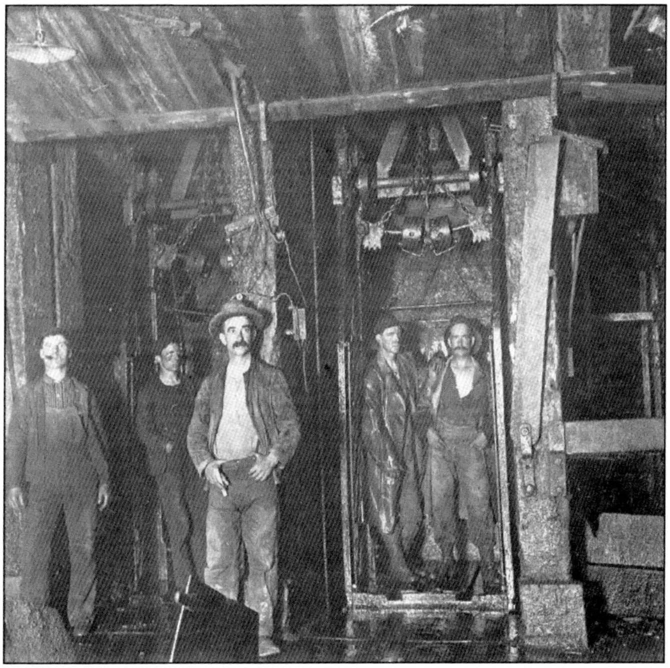

Shaft Station Deep in Montana Mine, c. 1900. The route from silver mine to silver dollar could be long and dangerous to the miners and laborers at every step. It is hard to understand the Neihart history without some knowledge of that route. Part of it involved working in dark, dank, and dangerous conditions. Even as miners advanced to using electrical mechanisms, such as the drills seen overhead in this Montana mine, operations and conditions did not necessarily advance at the same rate. (MOR.)

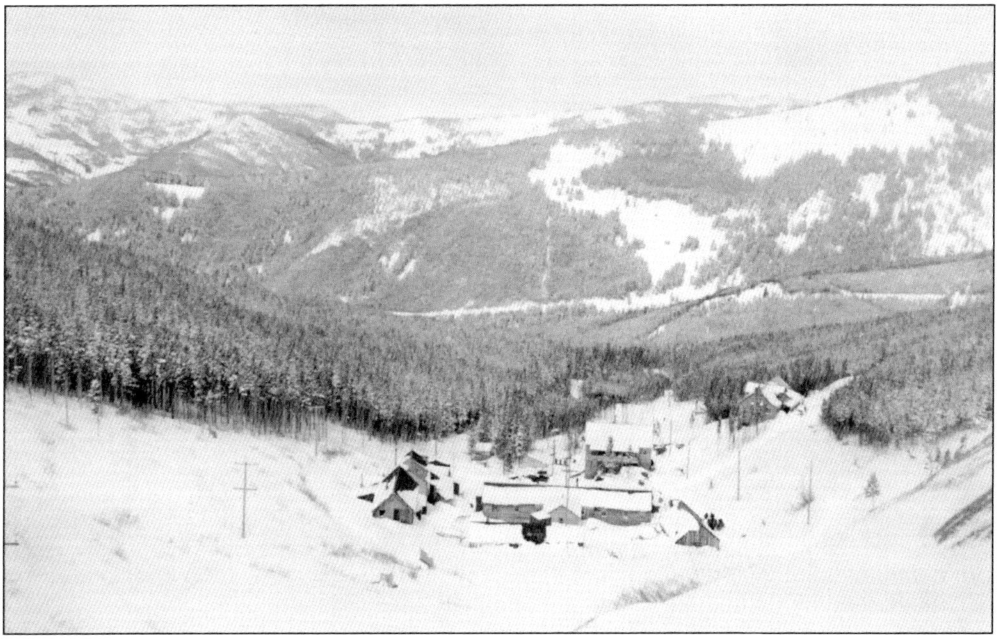

Lexington Mining Camp on Snow Creek. The Lexington Mine operations on Snow Creek are the subject of this winter photograph. In the early years, most miners only worked their claims from spring through early fall. They would then gather what belongings they required and wintered in Barker, Fort Benton, or White Sulphur Springs, where the snow did not accumulate quite as much as in the higher elevations and where they could avoid the temperatures of minus 40 degrees, which were not uncommon in Neihart. Once the mines began producing, and miners were able to work inside their mines during the winter months, more miners and laborers began constructing permanent homes, as seen in this photograph. (THM.)

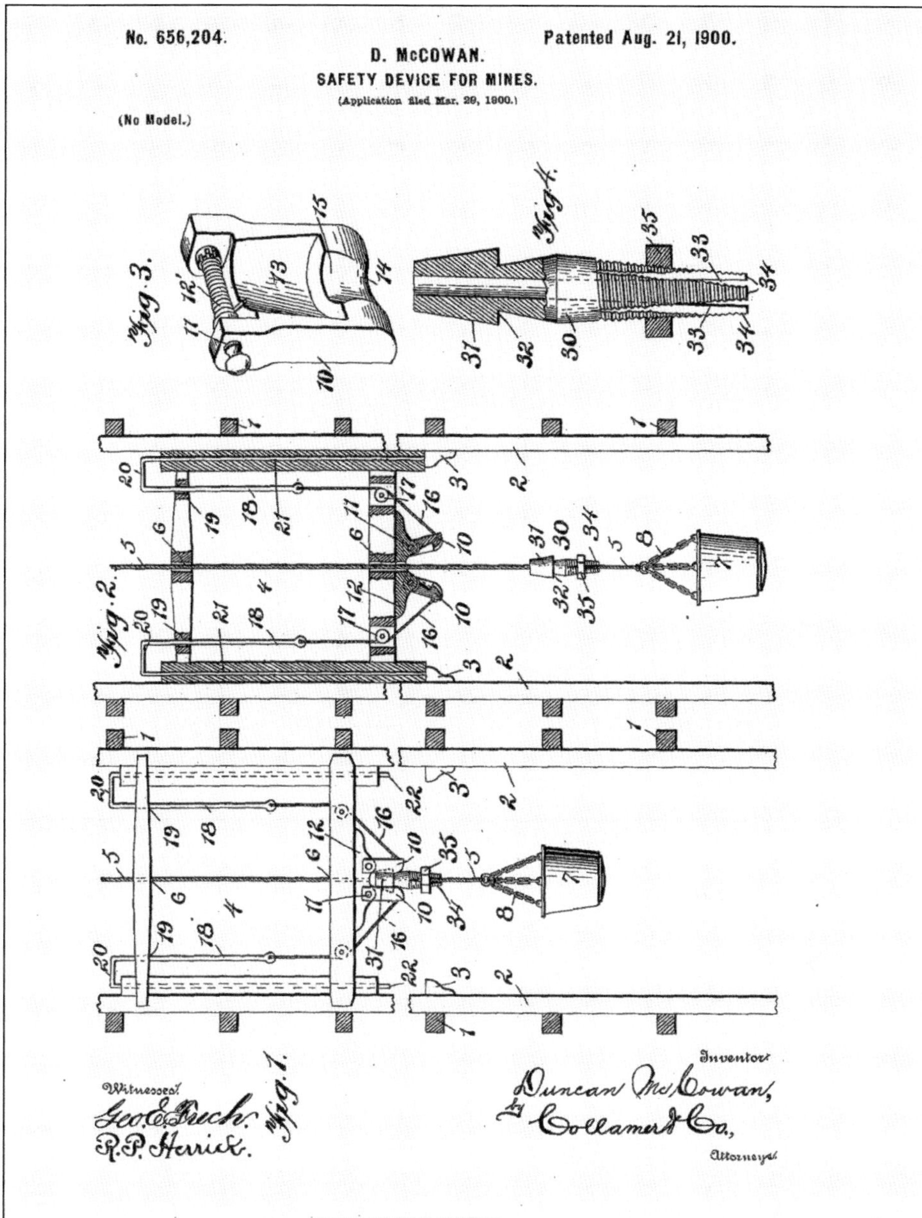

SAFETY DEVICE FOR MINES. Attorney Richard Bennett was not only a good land attorney and town attorney, he apparently was also a decent patent attorney. He filed a patent application for one of Neihart's citizens and miners, Duncan McCowan. McCowan invented a very important safety device for his fellow miners. The new self-arresting clutch was designed to prevent buckets and cages from self-releasing while under tension, which would cause the structure from which they were suspended to collapse. This device protected miners down below and prevented damage to the timbering in the shaft. Pictured above is the required drawing of the invention that accompanied the application submitted to the US Patent Office in Washington, DC. The application was witnessed by Joseph J. Schwerdt and Richard Bennett. Duncan McCowan, along with William Mueller, owned the Cumberland Mine. (Author's collection.)

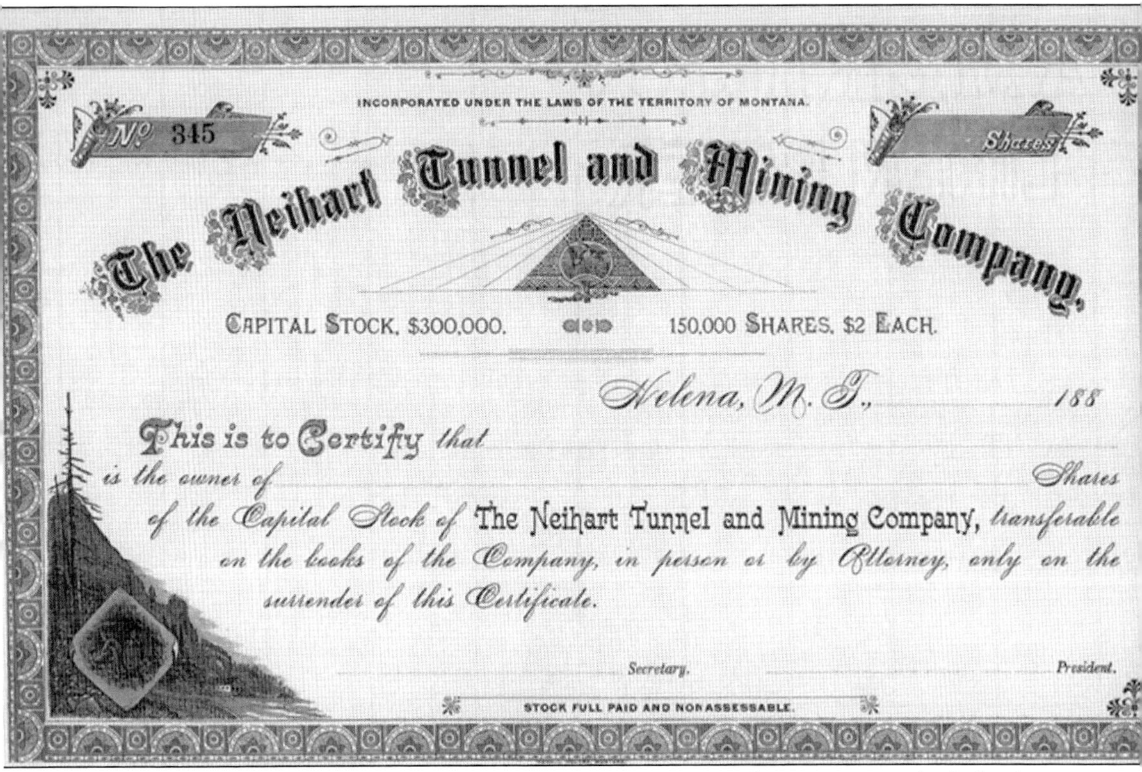

A COPY OF NEIHART TUNNEL AND MINING OWNERSHIP. Neihart was once the hub of activity for other mining camps in the Little Belt Mountains of Montana. With silver being discovered in 1881, the mining activity over the next 50 years fluctuated widely, dependent on the price of silver. Due to the remoteness of the area, transportation of ore was problematic. However, this did not deter the spirit or steadfast belief the miners held. They all wanted to issue certificates of ownership in their discovery as shown here regarding the Neihart Tunnel and Mining Company. Speaking of tunnels: there are more than six miles of tunnels coursing through the hills and cliffs that surround Neihart. (Author's collection.)

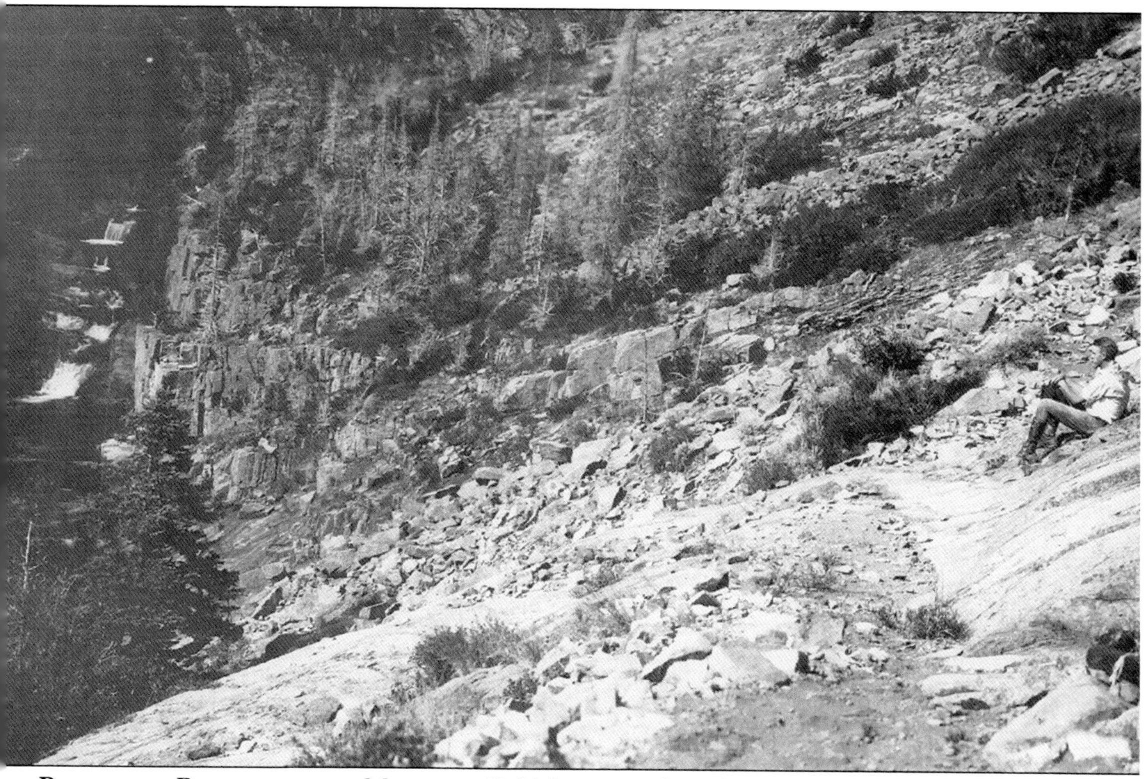

Beauty and Ruggedness and Minerals. Gold discoveries brought prospectors full of hope and high energy to Montana Territory in the 1860s. Initially panning for gold with some success, it resulted in searches of the entire mountainous region of the territory for mineral deposits. Not until 1876, when silver-ore deposits were found in Barker, did the Little Belt Mountains receive any type of mining activity. A few years following the Barker-Hughesville discoveries, three prospectors headed south up the gulch to find promising outcrops that Neihart was "discovered." The beauty and topography shown in this photograph are an excellent depiction of what these mountains hold. One can just make out the prospector at the right taking in the beauty. (MHS.)

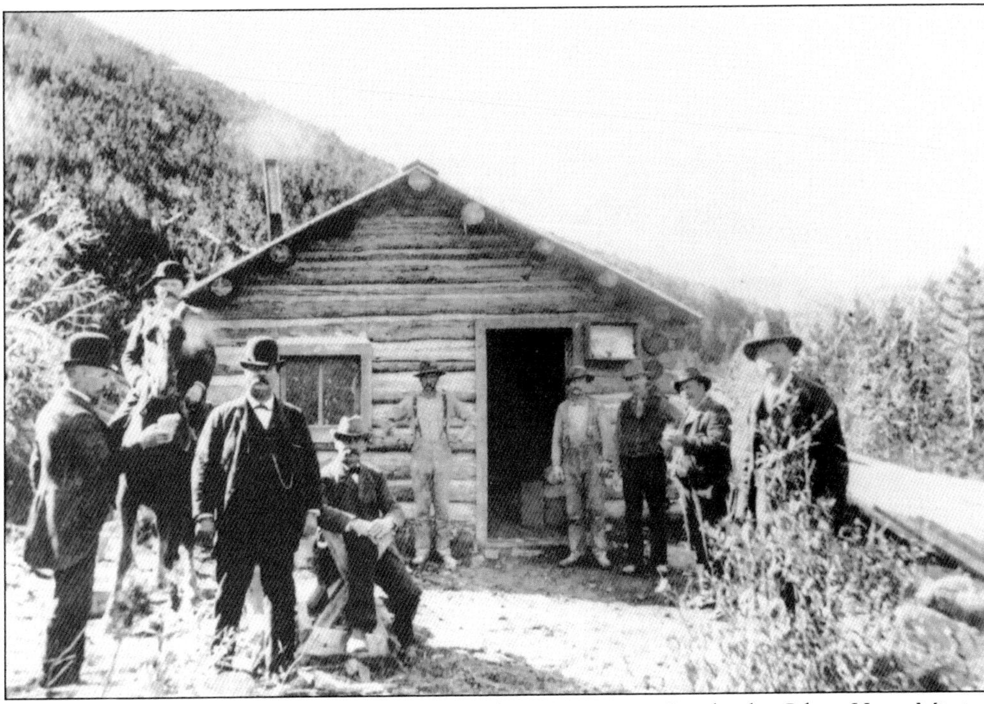

The Silver Horn Mining Company. Located on Carpenter Creek, the Silver Horn Mining Company claimed and worked three veins. The principal ownership consisted of P.M. Thompson (of Minneapolis, Minnesota) and Allen Pierse, Richard Bennett, T.E. Healey, William Tierney, Thomas Rothwell, John Hagerty, and Charles Black (all of Neihart). In the closeup, outside the Silver Horn office, Richard Bennett stands to the far left. In the photograph below, showing the owners at an adit, Bennett is seated third from the right. (Both, THM.)

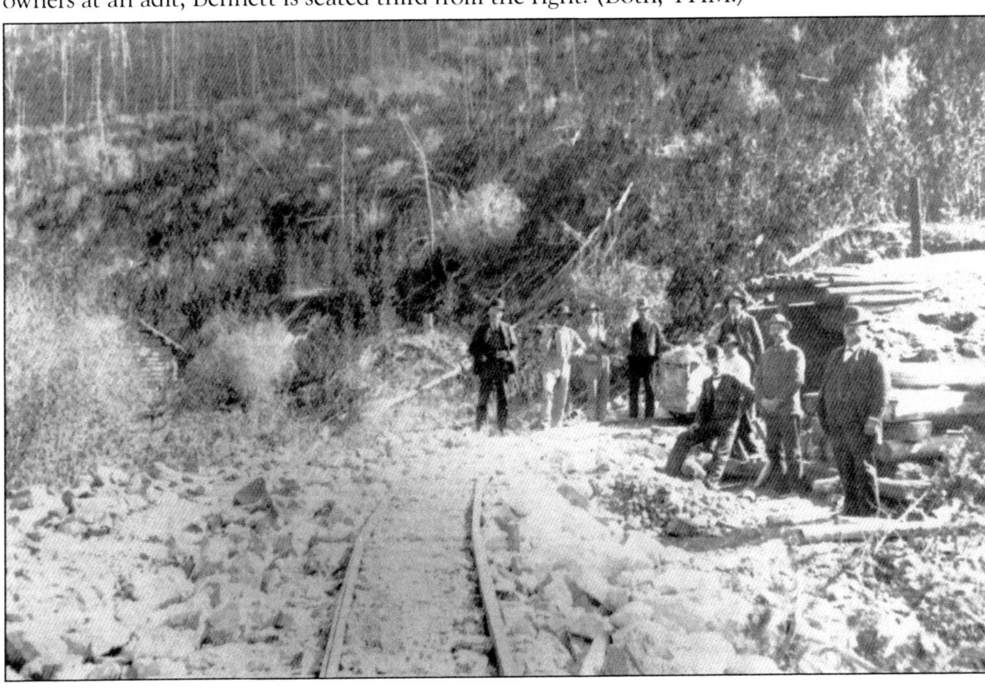

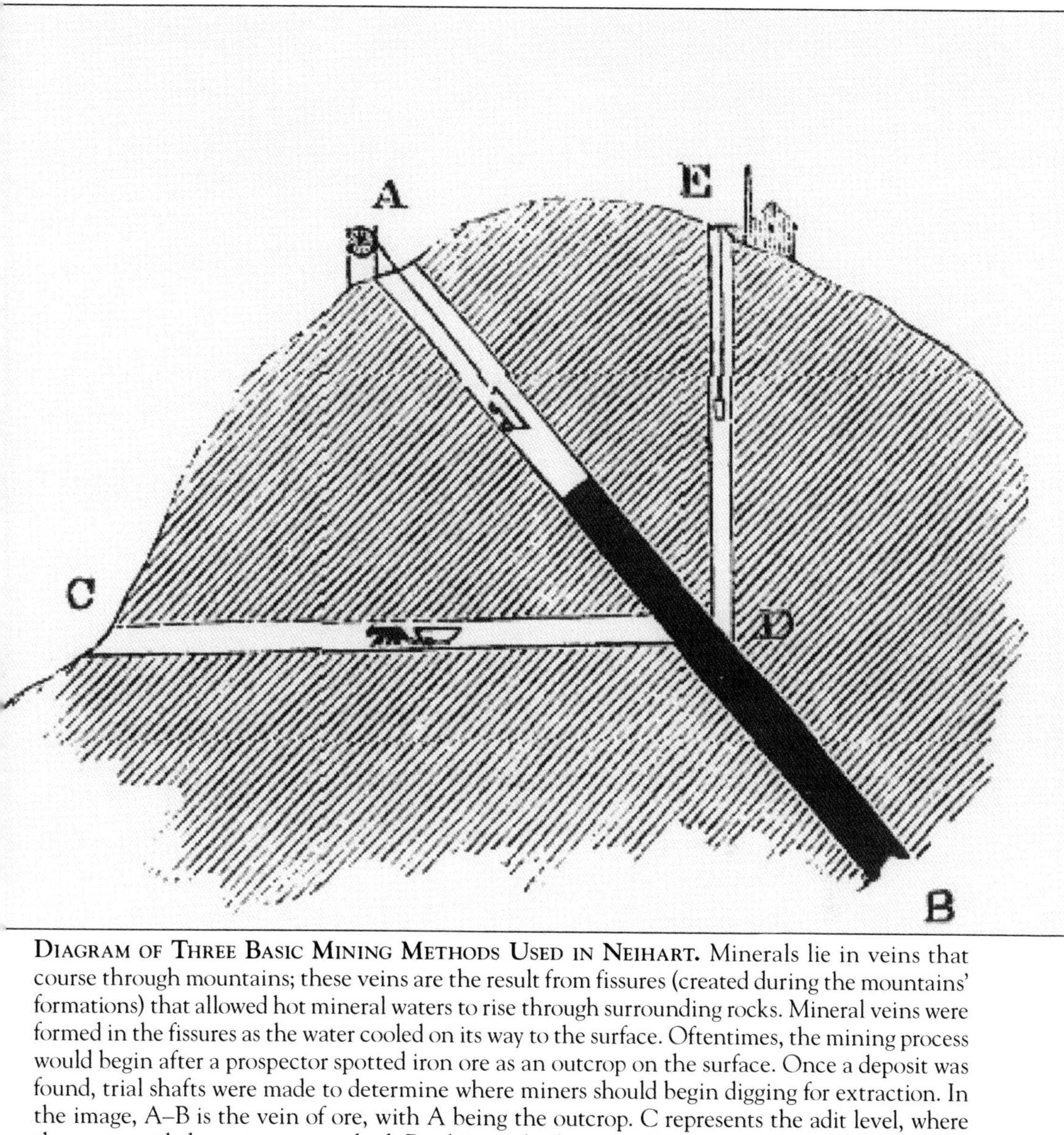

DIAGRAM OF THREE BASIC MINING METHODS USED IN NEIHART. Minerals lie in veins that course through mountains; these veins are the result from fissures (created during the mountains' formations) that allowed hot mineral waters to rise through surrounding rocks. Mineral veins were formed in the fissures as the water cooled on its way to the surface. Oftentimes, the mining process would begin after a prospector spotted iron ore as an outcrop on the surface. Once a deposit was found, trial shafts were made to determine where miners should begin digging for extraction. In the image, A–B is the vein of ore, with A being the outcrop. C represents the adit level, where they cut until the vein was reached. By this method, miners began removing the ore via the adit tunnel. Then galleries (like large rooms) were cut to the right and left, removing all the ore between D and A. A second method was to sink an inclined shaft on the vein. Following the vein and making galleries as they went, miners removed the ore using ore cars. A third method involved sinking mine shaft E and following the same process. This option was usually more expensive, with lower returns. (Author's collection.)

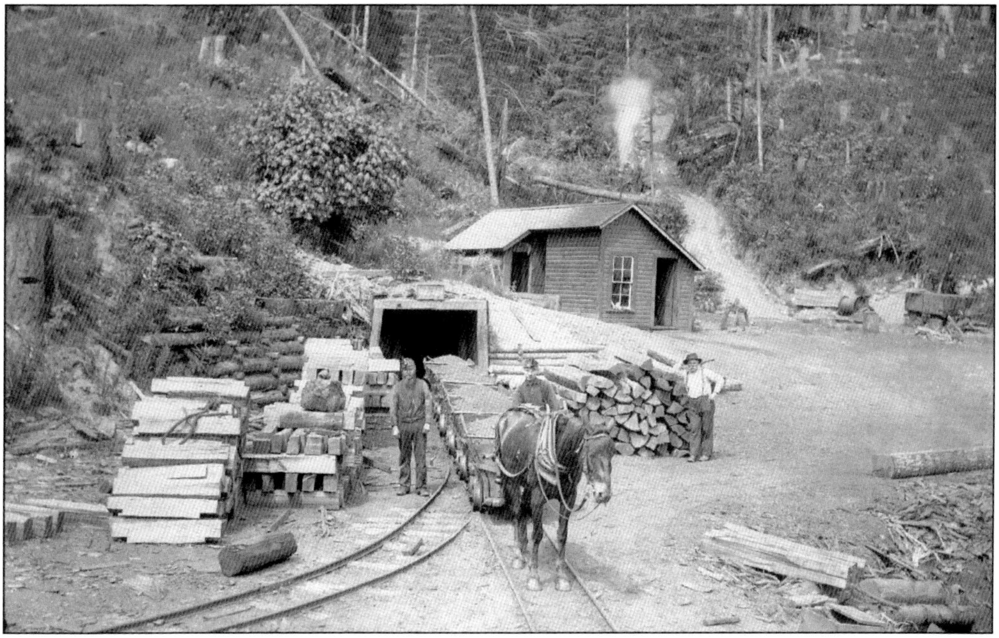

Horsepower Powers Some Mines in Neihart, c. 1883–1894. Most silver mines in the Montana Mining District (an official name for the Neihart area) used mules to remove ore from mines. This unidentified mine, in which an interest was held by James J. Hill of Saint Paul, Minnesota, who founded the Great Northern Railroad, apparently had a need for actual horsepower to pull ore cars to be transported, via the railroad, to Great Falls for processing. (James J. Hill Papers, MnHS.)

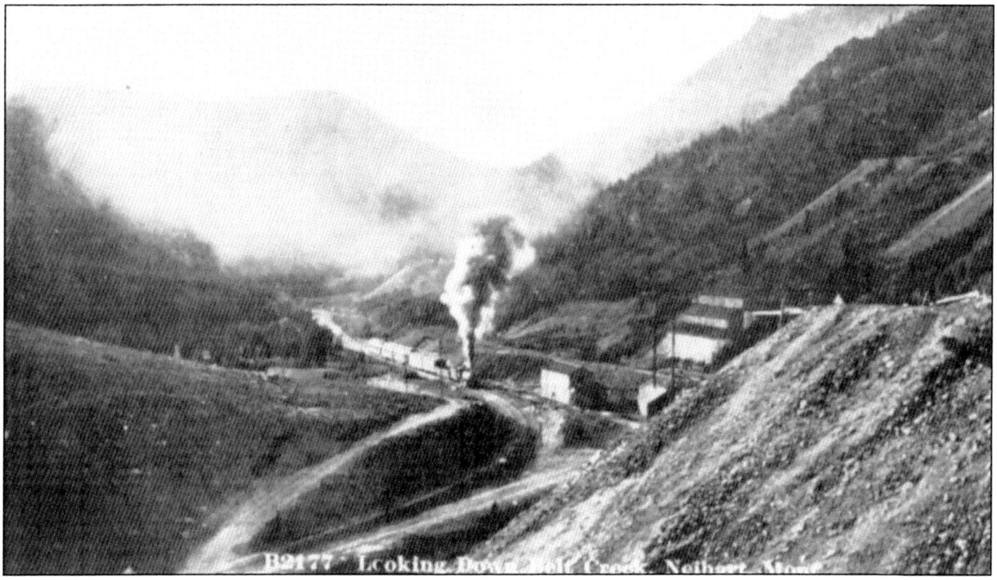

Montana Central Railroad Passing the Florence Mine. This photograph depicts the Neihart Flier passing the Florence Mine workings, downstream of the Queen of the Hill Mine and Neihart. (Tony Juntunen.)

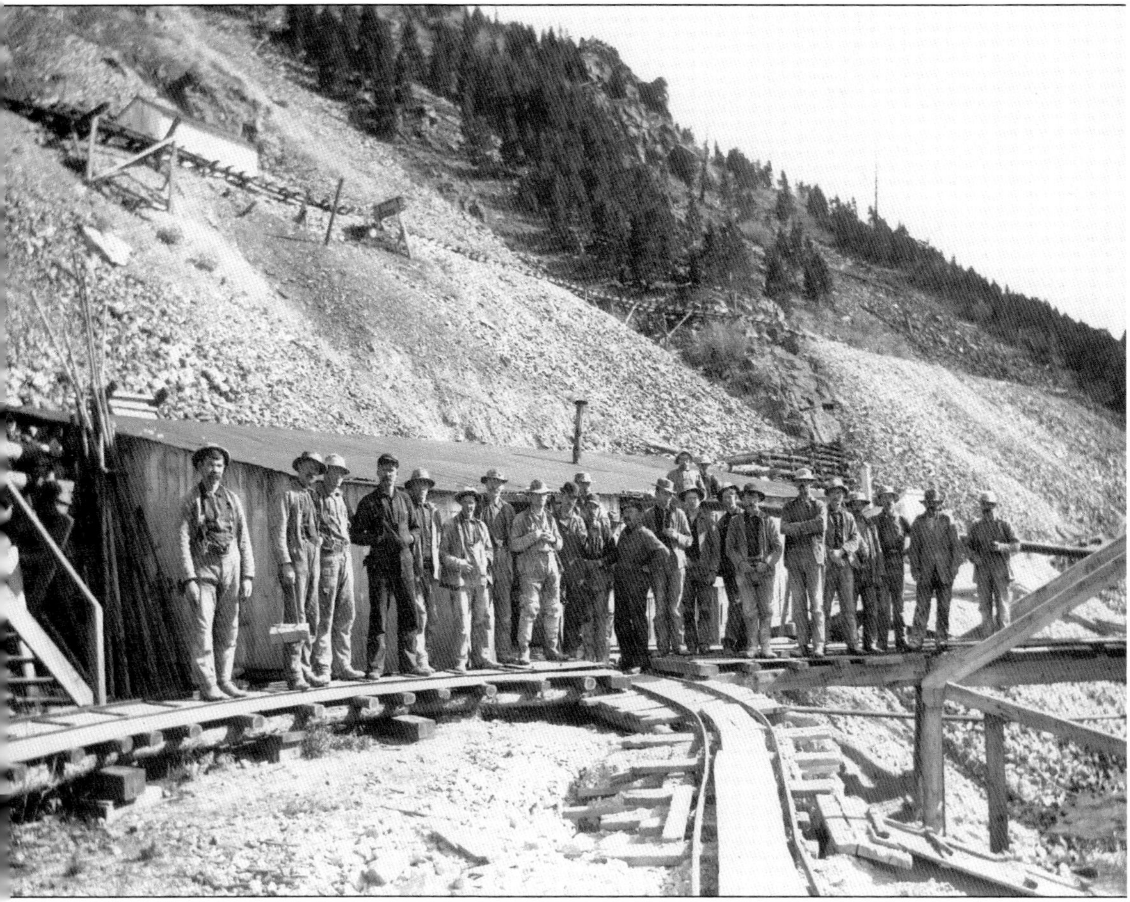

Florence Mine Men. This image provides a good view of the Florence Mine and the surrounding work. In the upper third of the photograph are ore car tracks, perhaps from another adit or tunnel. The tracks are supported across the gully by a trestle. The photograph also shows the extent of trackage laid, steps made, and braces shored. (MHS.)

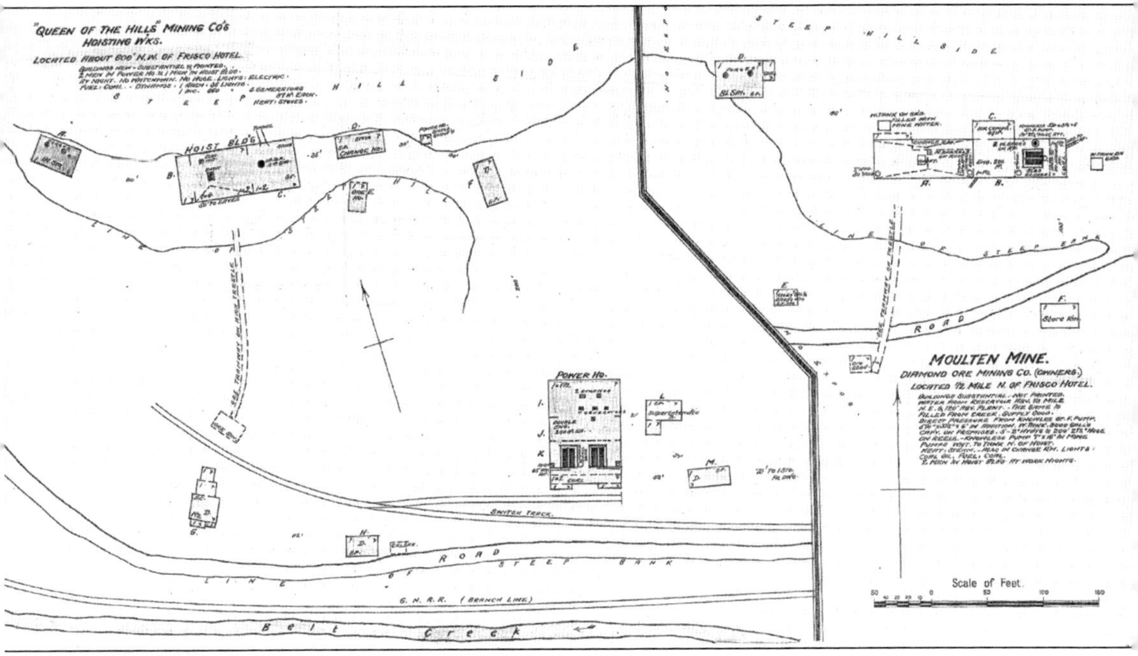

SANBORN FIRE INSURANCE PROPERTY REPORT. This image shows the physical layout of the Queen of the Hills Mining Company's hoisting works. It was drawn so an insurance company's underwriter could determine whether the mine owners were properly and wisely protecting their property from a known loss, as well as to document the safety items in place. As the inspector described, the mine was "located 600 [feet north west] of Frisco Hotel" and the buildings were "new [and] substantially painted." The report also documents the number of men on duty at night, the lack of watchmen and hoses, and the numbers and types of lights, generators, and other equipment. Such reports show how fire insurance rates were calculated and premiums determined across the country, not just on mines in Montana. (MHS.)

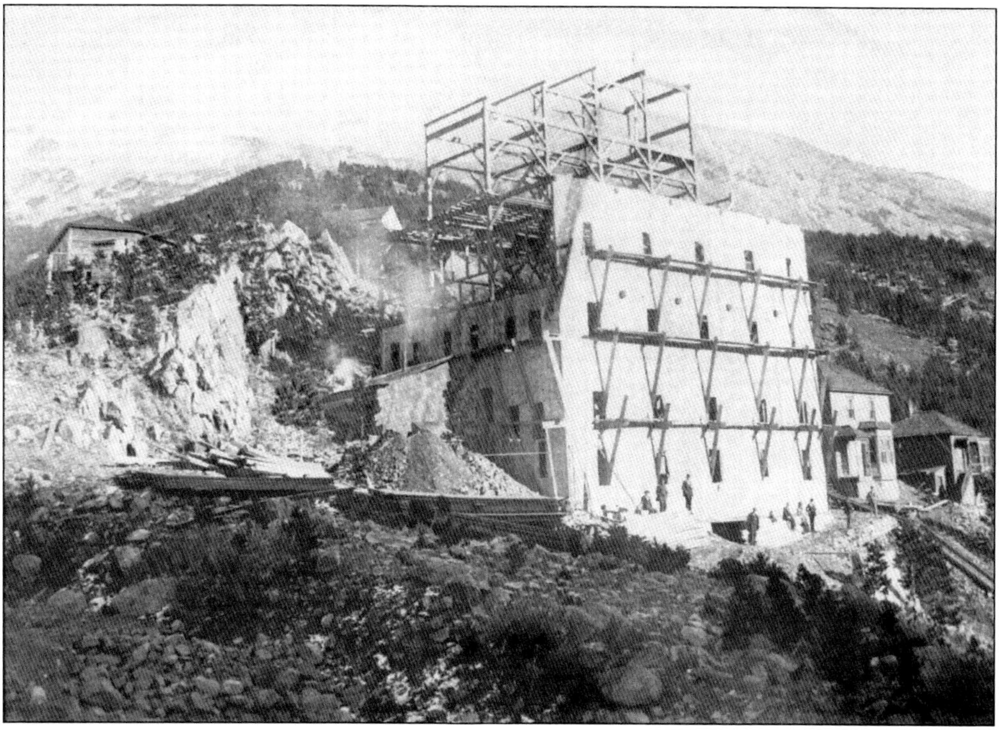

DIAMOND R COMPANY CONTINUED PROMOTING OTHER MINING WORKS, C. 1900. The Diamond R Company's mines worked a vein of sulphide ore thought to be the "finest ever found." The ore, three to five feet in width, was mined using compressed air drills and assayed at a high value, which was common for most mines sitting low on Baldy Mountain. To help treat lower-graded silver ore for their Moulton Mine (and others), Diamond R Company installed a 150-ton capacity concentrator, creating a system to increase profits for all. The concentrator's ability to save a high percentage of the minerals at a small cost was critical for this area. According to John Byrne, Montana inspector of state mines (1889–1890), "the future progress and development of the Neihart district depends altogether on the enterprise displayed in erecting additional mills." Simply stated, the other miners need this mill for their success, too. (Both, MSH.)

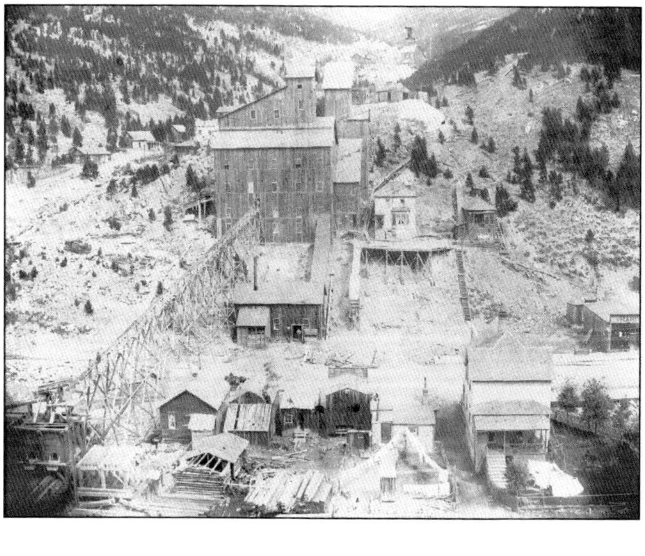

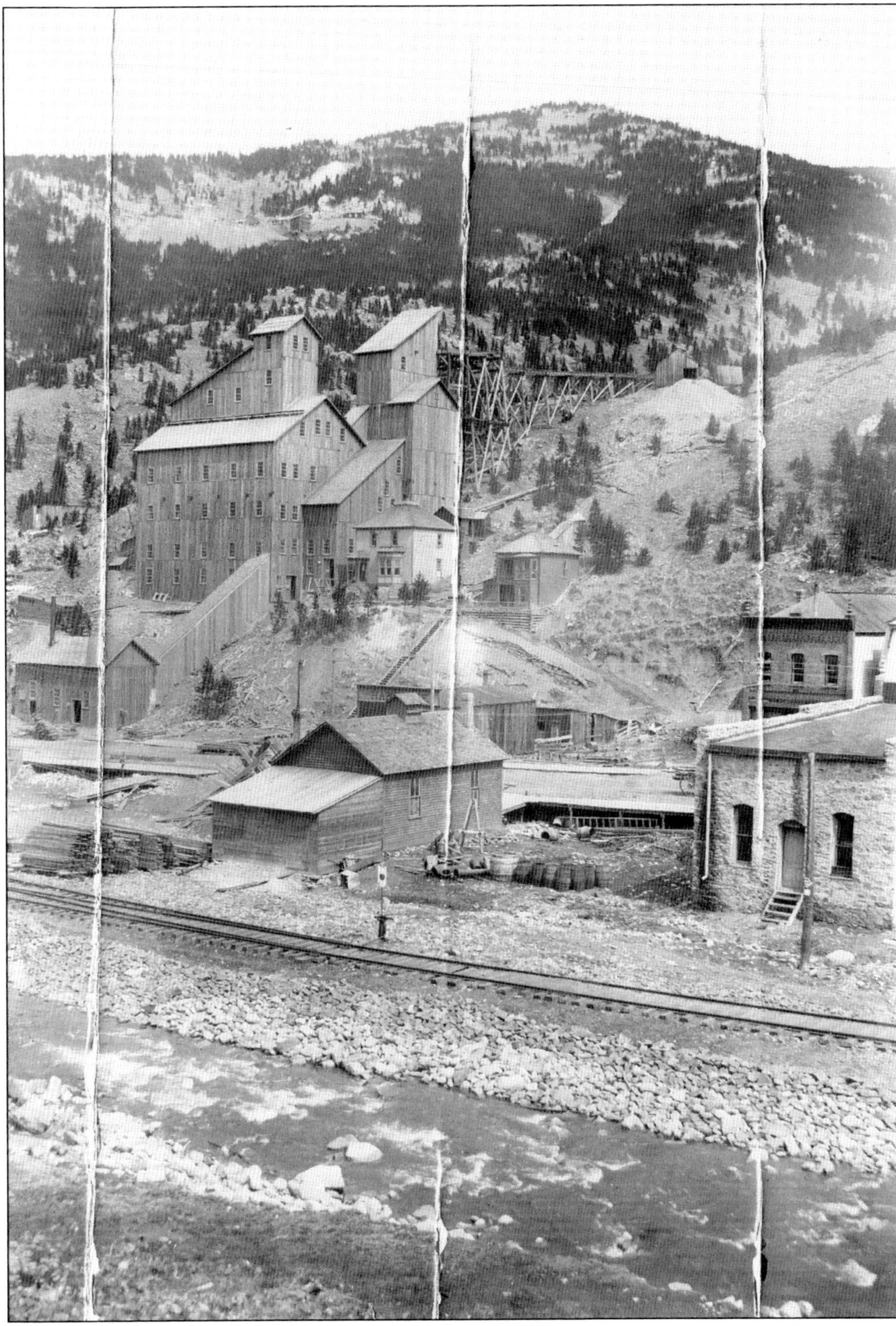

DIAMOND R TRAMWAY. Another progressive development was the construction of a tramway from the Broadwater group of mines on Baldy to the Diamond R Concentrator. Above Neihart, a few of the head frames for that tramway system are still visible today. (THM.)

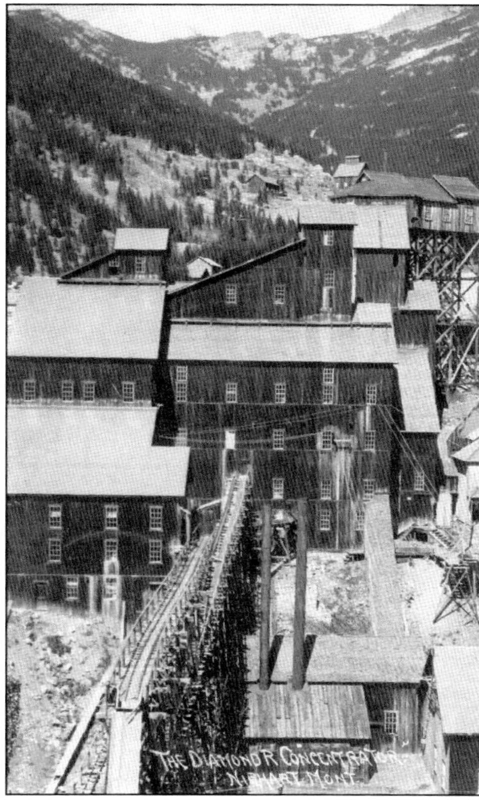

The Diamond R Concentrator Commanded a Large Footprint in Camp, c. 1900. The Diamond R Concentrator filled a large portion of the Neihart skyline. Even by today's standards, the Diamond R Mining Company built big and tall. The concentrator was visible from almost anywhere in the upper canyon, and other structures on the banks of Belt Creek, such as the Florence Mine and the Queen of the Hills workings, made for an impressive scene. The Diamond R Mining Company was owned and operated by E.G. Maclay, Ira Myers, and A.G. Kingsbury, all of Great Falls. C.D. McClure of Saint Louis (with some silent partners) eventually became the managing partner of operations. (Both, THM.)

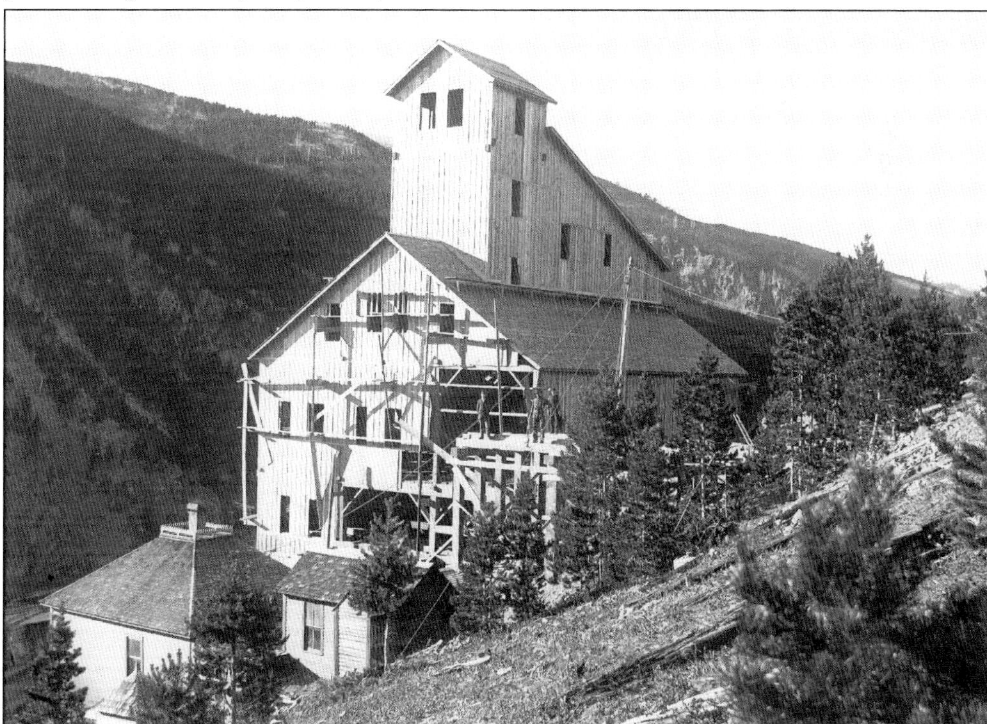

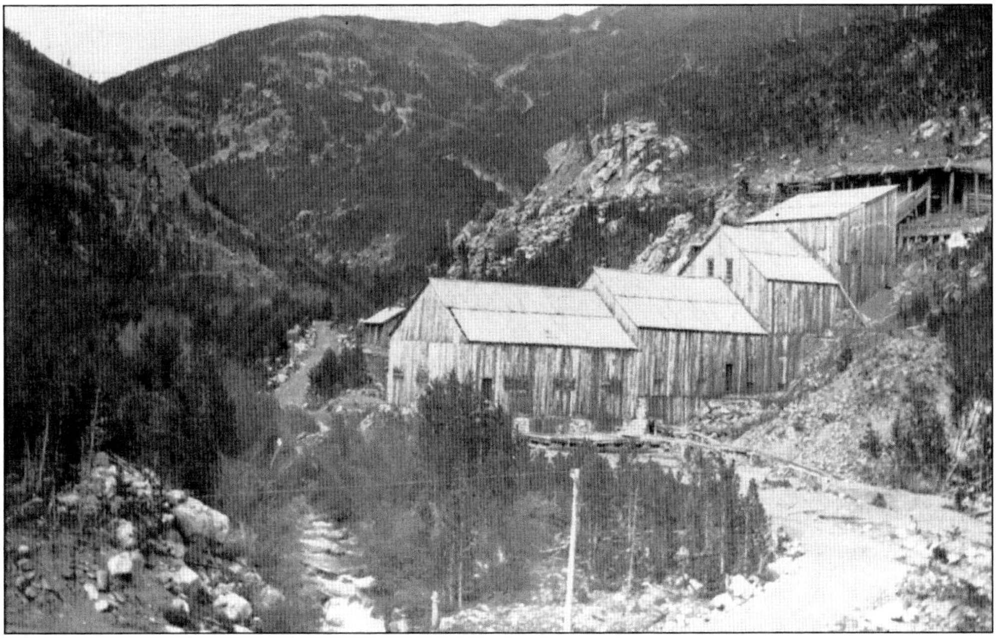

Northward View of Carpenter Creek from Belt Creek. From the confluence of the two streams, Hudson Mining Company's Concentrator is located to the right. J.C.E. Barker came into Neihart with his uncle A. Caldwell in December 1883. Around April 1, 1884, Barker bonded the Hudson Mining Company with Boston capitalists, who were very interested in the area's silver production. (THM.)

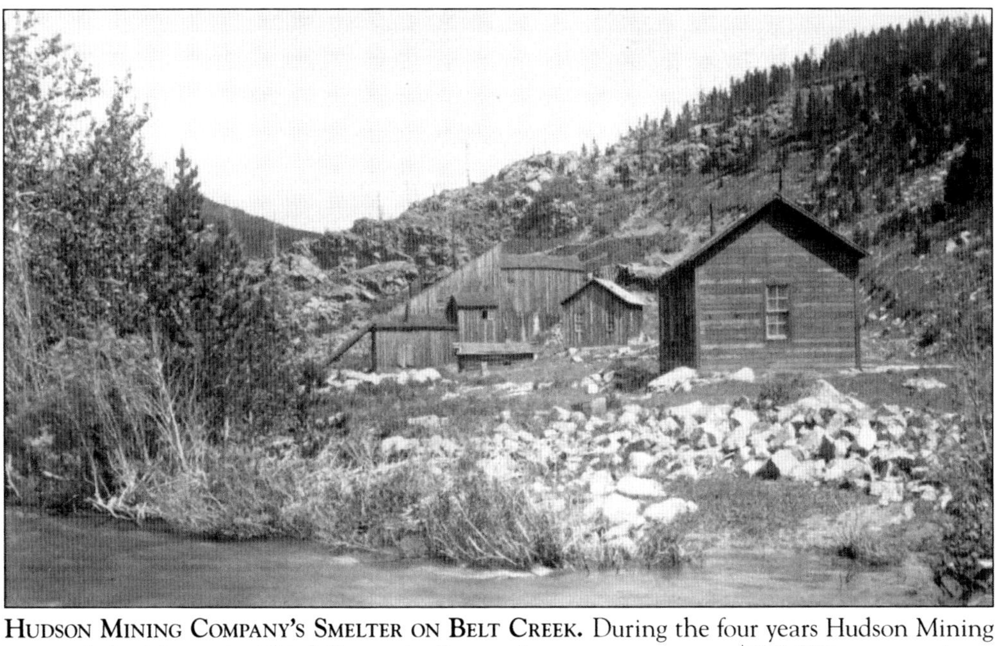

Hudson Mining Company's Smelter on Belt Creek. During the four years Hudson Mining operated the Mountain Chief Group, its Boston investors spent over $200,000 constructing a mile-long flume, concentrator, and the smelter pictured here. The total production for the group has been difficult to determine but suffice to say, after the first crash of the price of silver in 1887, the capitalists from Boston soon lost their appetite for Montana mine speculation. (THM.)

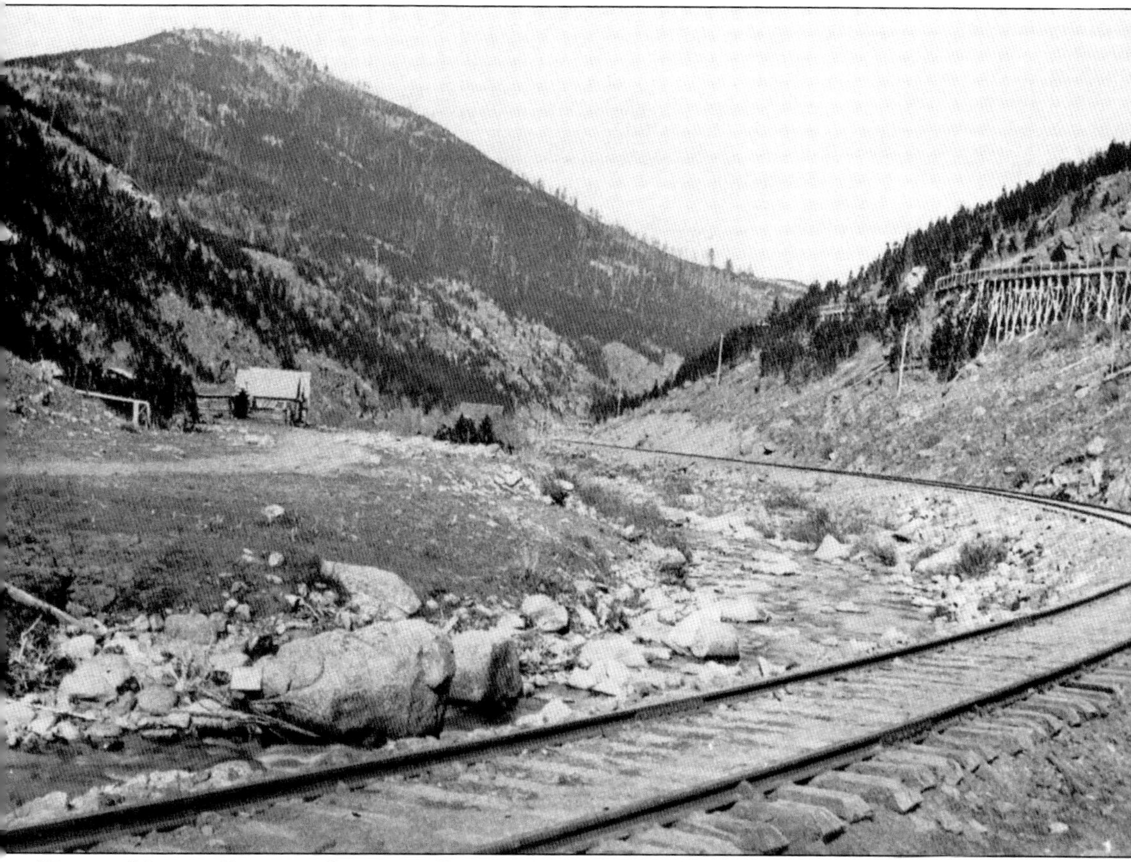

Hudson Mining Company Owned by Boston Capitalists. The Mountain Chief Group, located on Carpenter Creek near the confluence with Belt Creek, operated between 1884 and 1887. In this picture, the flume wraps in and around the hill side, stretching out in the distance around the farthest bend to the concentrator. J.C.E. Barker of Great Falls was superintendent while T.C. Simpson of Boston represented the Boston interests. Following the discovery and operation of the Mountain Chief was the establishment of the mining camp known as Jericho, which was about two miles north of Neihart. Jericho once entertained ideas of overtaking Neihart, but just the opposite occurred. Neihart included Jericho into its corporate limits when Neihart was incorporated by the county of Meagher. The mine was reported to have driven a tunnel 1,700 feet in length with two shorter tunnels running above on the ore lode. (THM.)

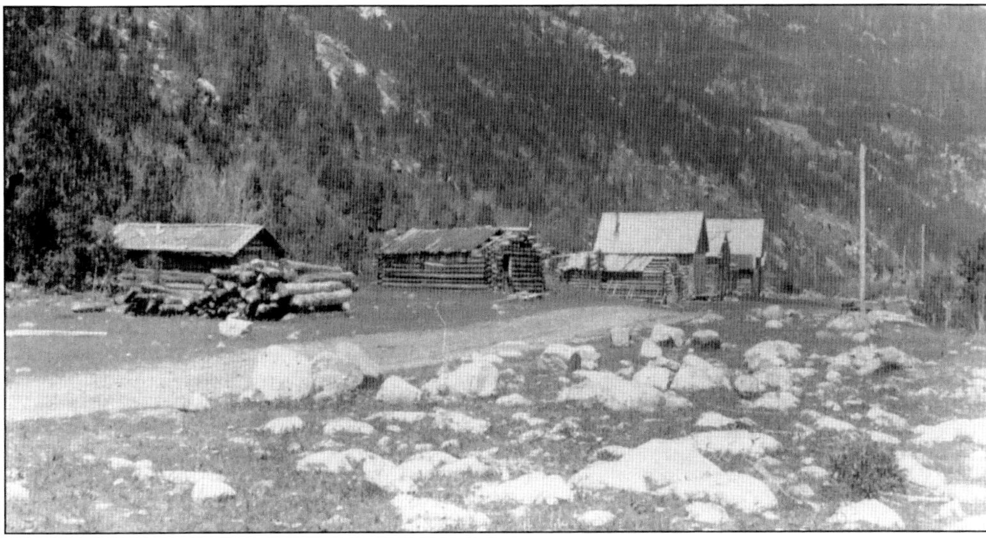

Jericho, a Suburb of Neihart. Notice the main road eventually going north to Monarch and south to Neihart. The Montana Central tracks ran where Highway 89 is now found. Jericho, at the height of operations, had about 80 residents. Sarah Neihart ran the boardinghouse, and the saloon was run by a Scotchman known as "Granny" or "Jericho Mack." The first log shack built as a result of the pending mining activities in the area was erected here by Ed Tingal. (THM.)

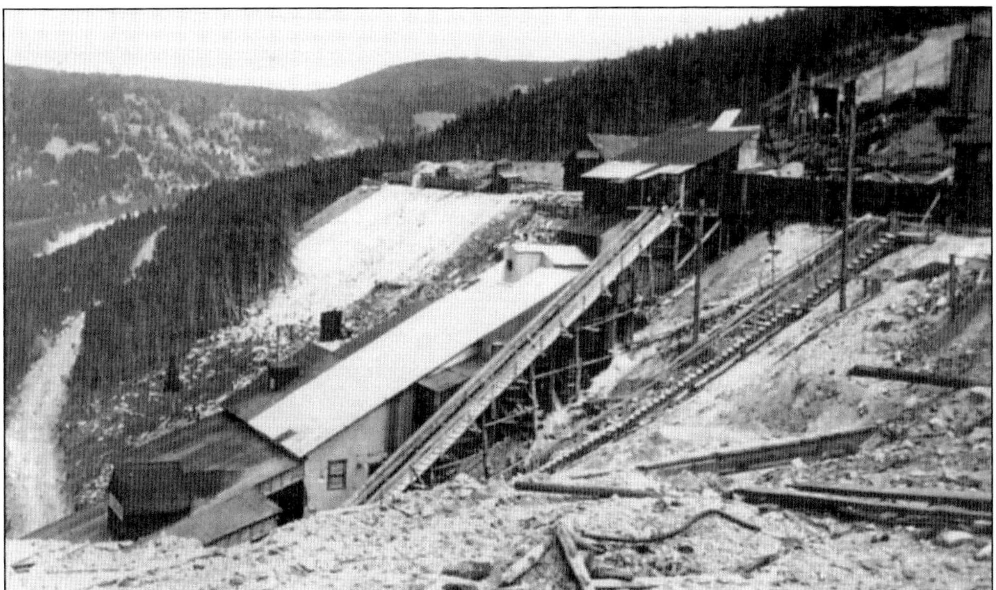

Flotation Mills. This particular flotation mill is located on either Snow Creek or Carpenter Creek in the Neihart District. Flotation mills were introduced around 1917, replacing concentrators and the pulverization of rock to separate metallic mineral from the ore. Flotation mills separated metalliferous (silver, gold) minerals from waste rock, using surfactants to cause air bubbles to attach to mineral particles and float to the top of a frothy bath of ore slurry (similar to the effect of dish soap and water on cooking grease). The goal of flotation mills was to maximize the amount of valuable material floated, while minimizing the ore concentrate's gangue (worthless rock) content. (THM.)

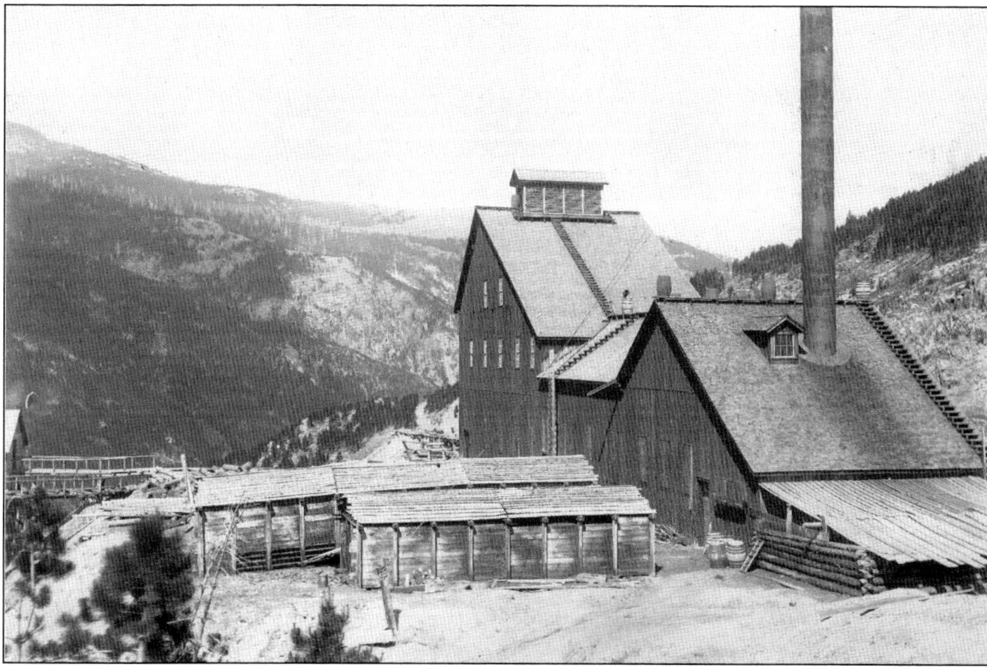

The Moulton Mine. Owned by the Diamond R Mining Company, the Moulton was located on Rock Creek within the Neihart mining district. Upon entering, one took the road up and around the Queen of the Hills towards Rock Creek before reaching Neihart's Main Street. Up Rock Creek, owners E.G. Maclay, Ira Myers, A.G. Kingsbury—all of Great Falls—and C.D. McClure of Saint Paul, Minnesota, operated the Moulton. This photograph faces west, downstream towards Neihart. (THM.)

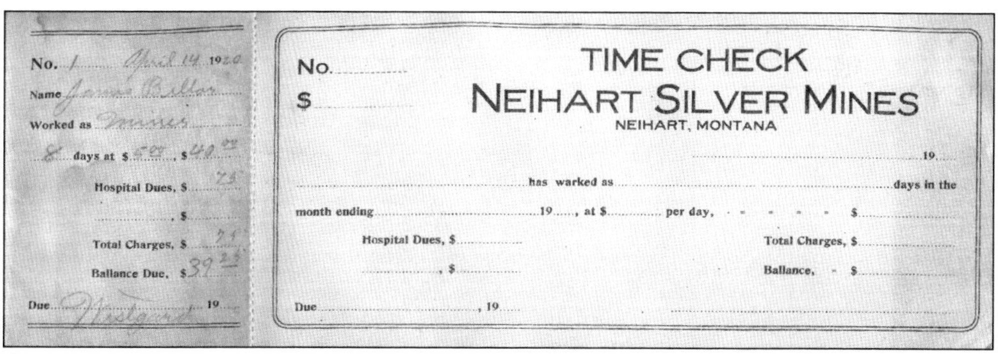

No. 1 April 14 1920
Name
Worked as Miner
8 days at $5.00 $40.00
Hospital Dues, $.75
, $
Total Charges, $.75
Ballance Due, $39.25
Due 19

No.
$
TIME CHECK
NEIHART SILVER MINES
NEIHART, MONTANA
19
has warked as days in the
month ending 19 , at $ per day, $
Hospital Dues, $ Total Charges, $
, $ Ballance, $
Due , 19

Time Check for Neihart Silver Mines. This time-check book reveals more about the period in which it was used than just the employee's pay check. This particular book was used only nine times, indicating that nine men received a time check to present to a paymaster or bookkeeper at the end of the day. This book illustrates what happened after 1918 when metal prices had another small boom. Life in Neihart picked up for a while with the rejuvenation of mine activity. This time-check book starts on April 14, 1920, following World War I and the little burst of mining activity. That day, a miner was issued a time check for eight days of work at a rate of $5 a day. Net pay, after a 75¢ deduction for hospital dues (Neihart's hospital illness and injury plan for miners), was $39.25 for eight days of work. Some earned $5.50 a day, so the going rate for working the mines at the start of the Roaring Twenties was not much to roar about in Montana. (Ruth Bach.)

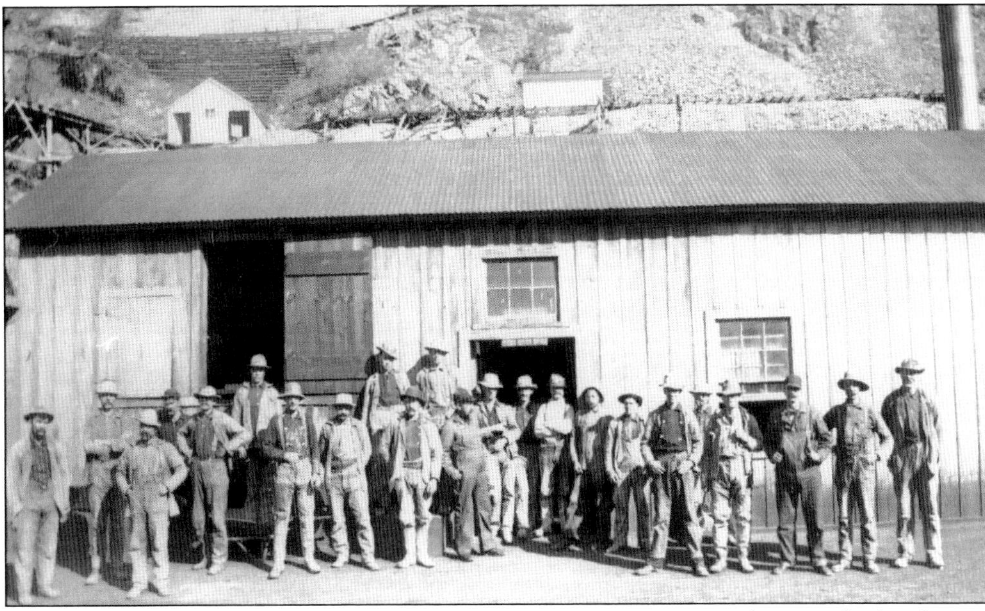

Miners of Broadwater Group. Unfortunately, this group of miners, photographed in front of one of the buildings of the Broadwater Group, cannot be identified. This picture was certainly taken before that eventful Saturday, April 23, 1896, when, at 1:05 a.m., havoc reigned over the Broadwater Group and Neihart. Powder had apparently been laid out to thaw while the miners were on their lunch break (12:00 a.m.). The *Belt Valley Times* speculated that, on their return, shortly after the foreman entered the area where the powder was stored, a candle in a sidewall fell onto the powder, setting off an explosion. Seven men died, and fifteen were injured. (THM.)

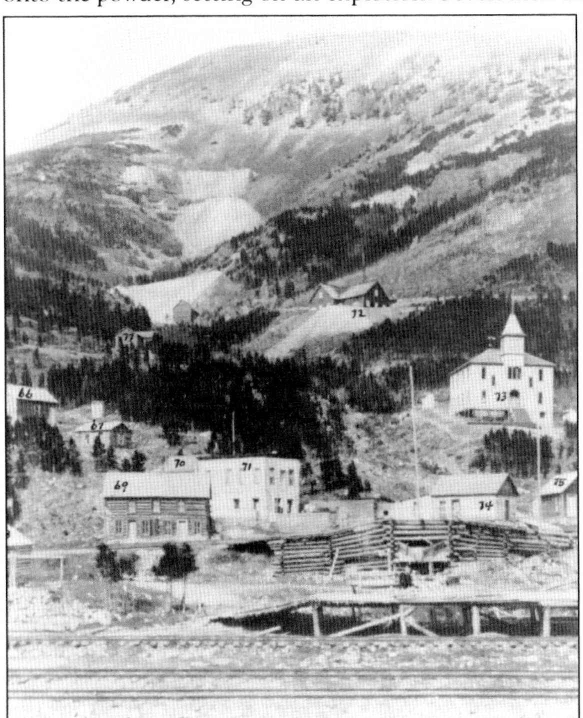

View of Broadwater Mining Operations on Neihart's Baldy, c. 1893. This photograph shows the railroad yard in the forefront and progresses to an excellent view of the land and elevation (8,286 feet) involved in mining the narrow Belt Gulch. Also shown are various residential structures in the middle of town. The identified structures in the image include the homes of Fred Hart (No. 66), C.H. Burchard (No. 69), John A. Workman (No. 71), Pete Rush and Henry Sutton (No. 74), Bill Spence (No. 75), and Aaron Johnson (No. 77). Also visible are the Lee Sing Chinese Laundry (No. 68), three Broadwater Group mine tunnels (No. 72), and Neihart School (No. 73). (THM.)

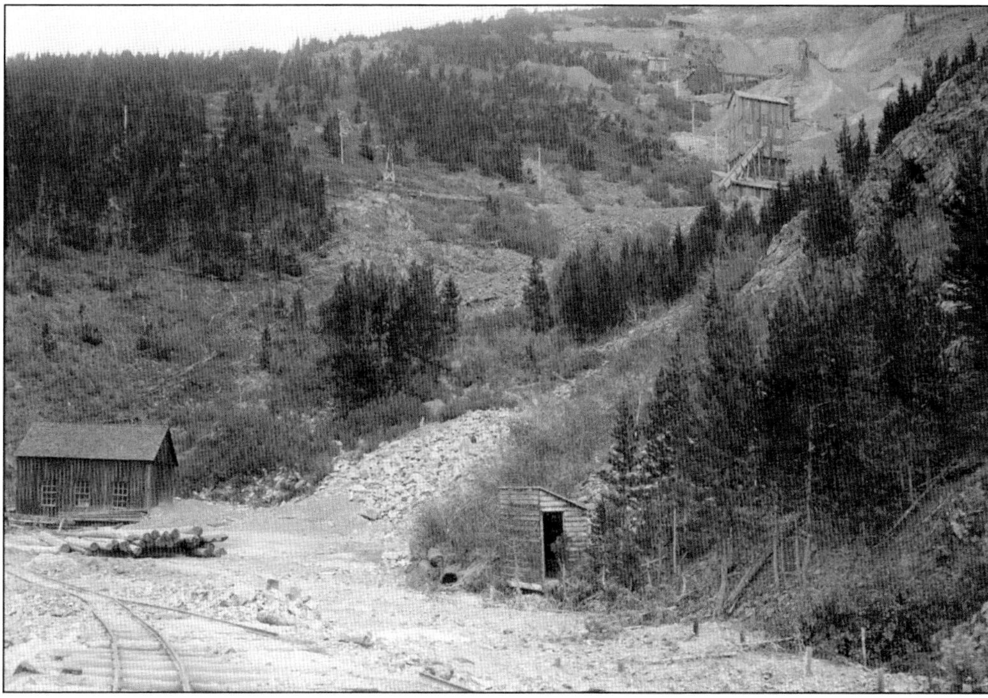

Broadwater Group Mining Operations. This group of mines—St. Julian, Minnehaha, Maud S, Montana Belle, and Dickens—lay on the west side of Baldy, roughly one-third of a mile above town. The image shows a number of structures that were key for the operation of these mines. Originally operated by C.A Broadwater in partnership with James J. Hill, the mines employed around 75 men. Upon Broadwater's death, W.J. Clark (of Copper Kings fame) and others purchased the operations for $165,000 in 1889. (MHS; photograph by E.R. Shepard.)

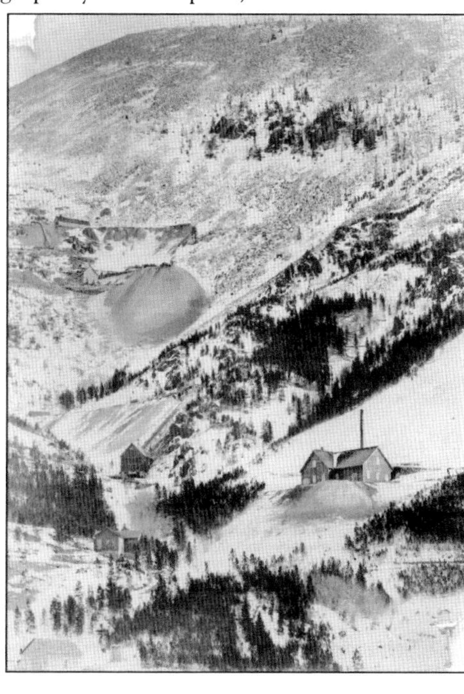

Broadwater Blast. On April 23, 1896, Neihart was rocked by an explosion that killed seven miners and injured fifteen others. W.J. Clark put all widowed spouses on the Broadwater payroll for a year and paid the injured miners until they could return to work. The *Belt Valley Times* reported that his "assistance in their hour of need [would] never be forgotten." The explosion occurred at 1:05 a.m., a result of an errant sidewall candle falling into a magazine of powder, which had been laid out to thaw while the miners had lunch. The seven killed were Frank Doran, Hugh McKenzie, D.H. Ross, John Cairns, Dan O'Leary, W.H. Morrison, and James T. Gallagher. (Author's collection.)

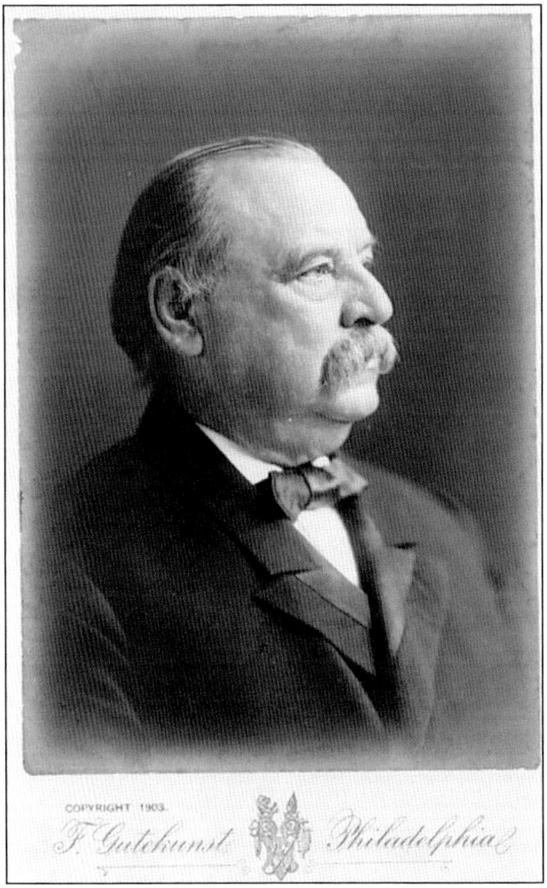

PRESIDENT OF THE UNITED STATES, GROVER CLEVELAND. A financial panic occurred in 1893, and many blamed the Sherman Act. Pres. Grover Cleveland convened a special session of Congress and demanded that the act be repealed. The silver-purchase and note-issuance provisions of the Sherman Act were in fact repealed in 1893, although the legal-tender status of silver coin and treasury notes remained. (Photographer, Gutekunst of Philadelphia, 1902; LOC.)

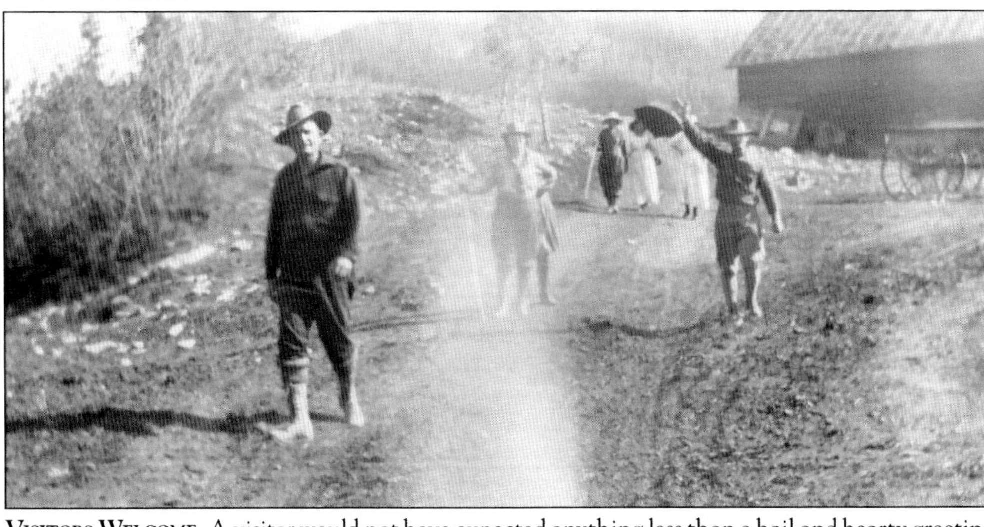

VISITORS WELCOME. A visitor would not have expected anything less than a hail and hearty greeting when arriving at any of the area farms or ranches in and around the Little Belt Mountains. That is exactly what is happening here as two unidentified men greet the photographer, with one or two unidentified women in the middle, slightly obscured, followed by three unidentified women enjoying their stroll with a parasol. (Photograph courtesy of the Mary Sohl Collection; THM.)

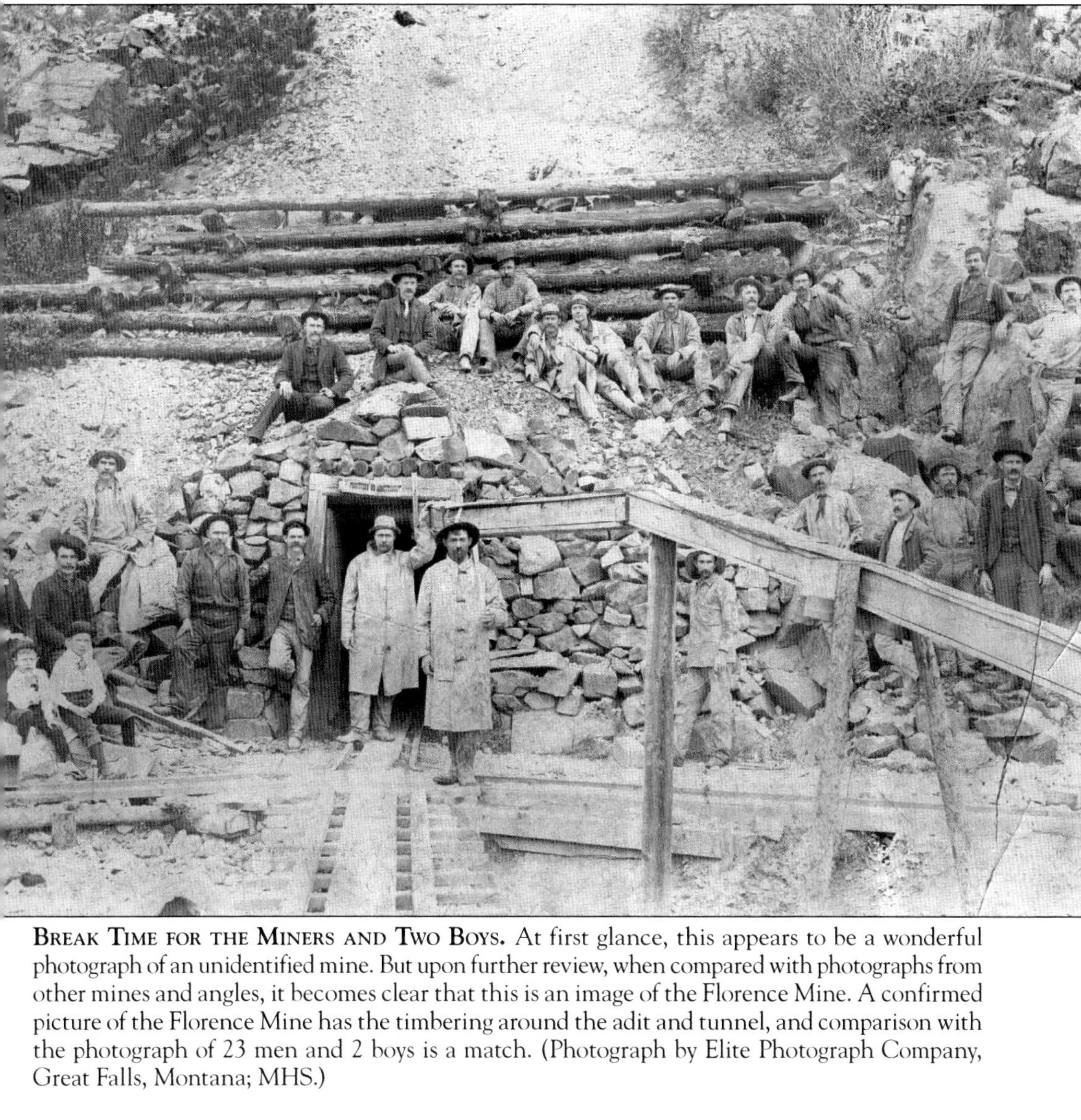

Break Time for the Miners and Two Boys. At first glance, this appears to be a wonderful photograph of an unidentified mine. But upon further review, when compared with photographs from other mines and angles, it becomes clear that this is an image of the Florence Mine. A confirmed picture of the Florence Mine has the timbering around the adit and tunnel, and comparison with the photograph of 23 men and 2 boys is a match. (Photograph by Elite Photograph Company, Great Falls, Montana; MHS.)

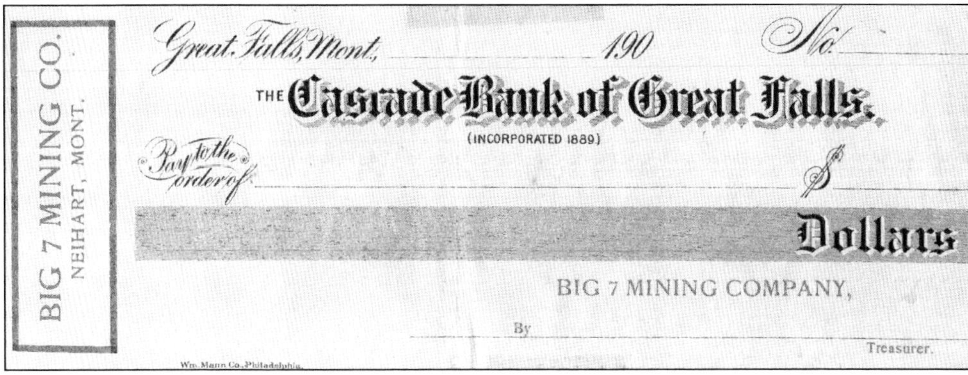
BIG 7 MINING CO.
NEIHART, MONT.
Great Falls, Mont, 190 No.
THE Cascade Bank of Great Falls.
(INCORPORATED 1889)
Pay to the order of $
Dollars
BIG 7 MINING COMPANY,
By
Treasurer.
Wm. Mann Co., Philadelphia.

Montana Miners Paid Well. According to the mine records of D.L.S Barker of Neihart, miners generally worked on contract, per tonnage, per foot or drift, or per cubic yard. Laborers who worked underground but did not do any actual extracting of ore were paid by the day. According to the *11th Census of the United States 1890* findings, Montana miners far exceeded the wages of miners across the country. A foreman aboveground made $5.39 a day while his cohorts elsewhere were receiving $4.04 a day. Underground foremen's wages were less: $4.16 per day for the balance of the country while in Montana Territory one would have received $5.11. For a laborer, it was the same: aboveground $4.14 per day in Montana and $3.67 outside of Montana. That difference in pay was most likely a result of getting and keeping good-working men. (Both, D.L.S. Barker Collection, THM.)

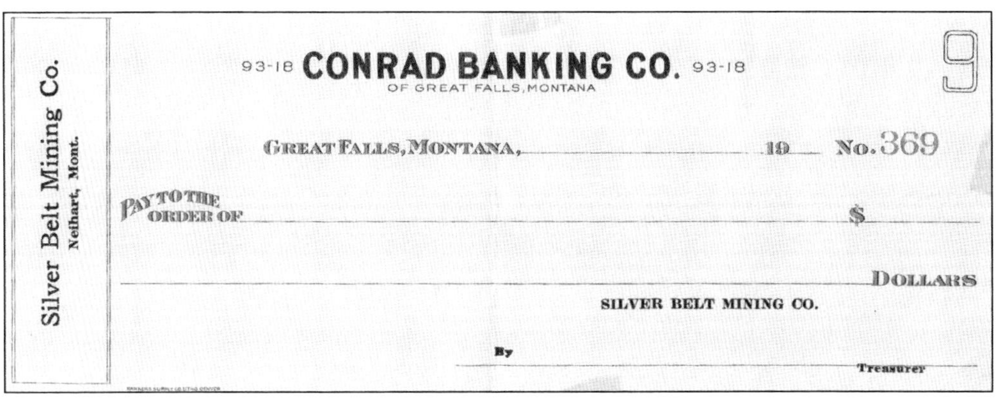
Silver Belt Mining Co.
Neihart, Mont.
93-18 CONRAD BANKING CO. 93-18
OF GREAT FALLS, MONTANA
9
GREAT FALLS, MONTANA, 19 No. 369
PAY TO THE ORDER OF $
DOLLARS
SILVER BELT MINING CO.
By
Treasurer

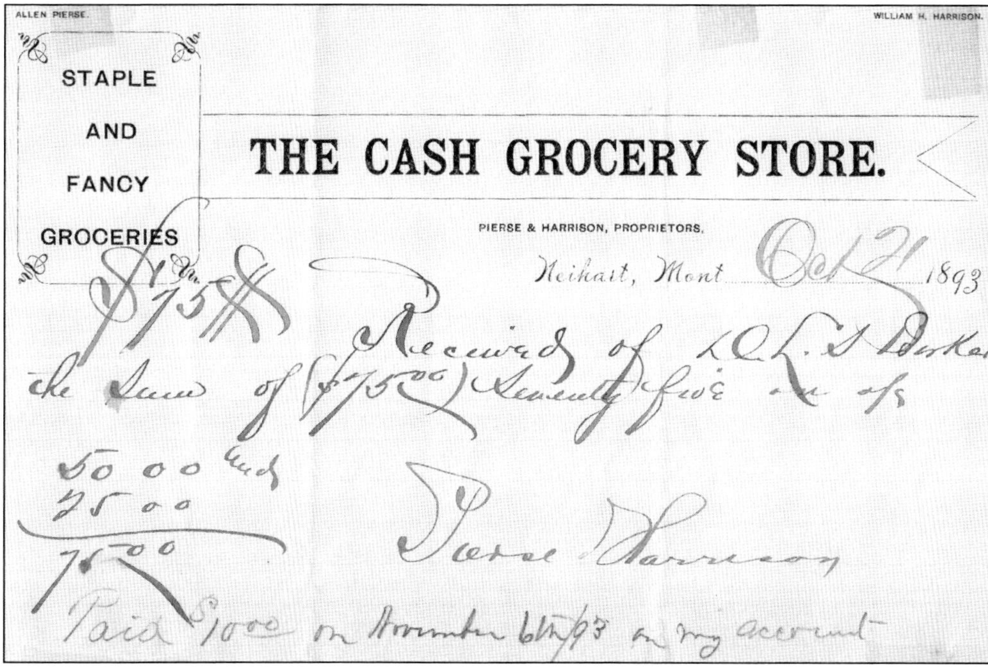

ALLEN PIERSE. WILLIAM H. HARRISON.

STAPLE AND FANCY GROCERIES

THE CASH GROCERY STORE.

PIERSE & HARRISON, PROPRIETORS.

Neihart, Mont Oct 21 1893

$75.00 Received of D.L.S. Barker the sum of ($75.00) Seventy five

50 00
25 00
75 00

Pierse & Harrison

Paid $10.00 on November 6th 93 on my account

Stocking Up on Supplies. While living and working in Neihart between 1893 and 1895, C.L. Barker, along with his brothers, took advantage of the merchant practice of granting credit in order to acquire food and mining supplies. (Both, D.L.S. Barker Collection, THM.)

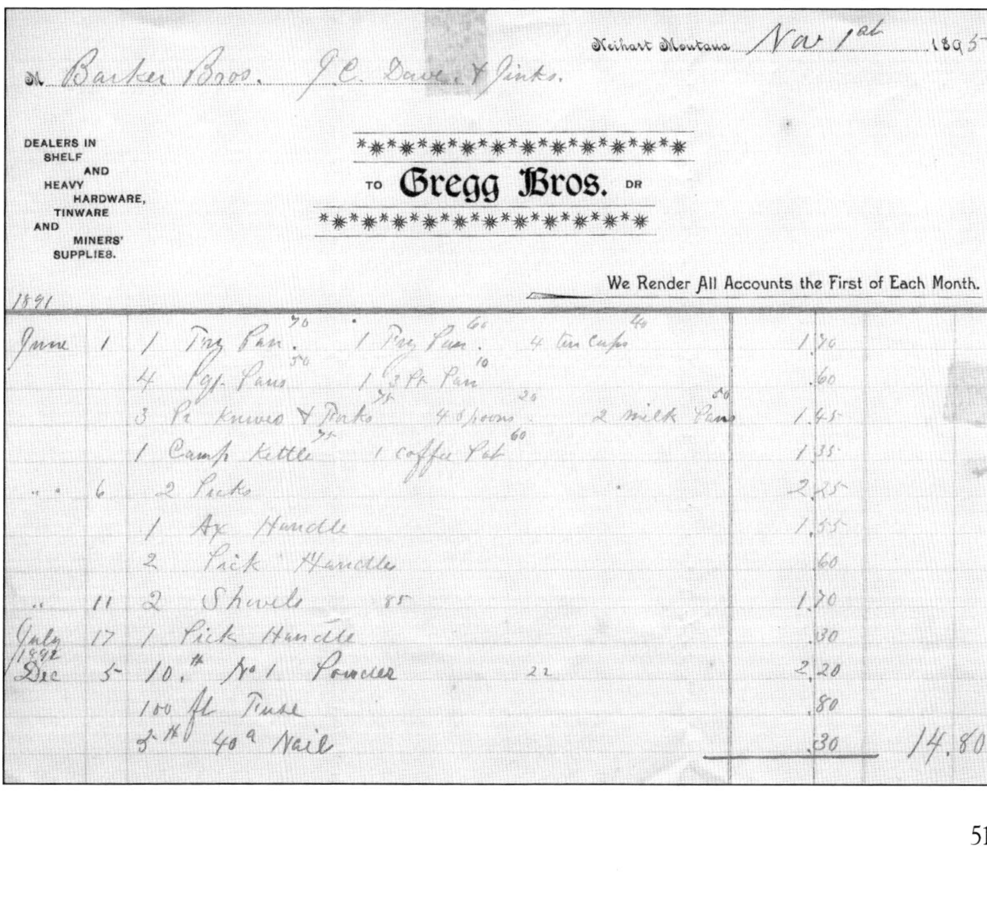

Neihart Montana Nov 1st 1895

M Barker Bros. J.C. Dave & Jinks.

DEALERS IN SHELF AND HEAVY HARDWARE, TINWARE AND MINERS' SUPPLIES.

TO Gregg Bros. DR

We Render All Accounts the First of Each Month.

1891				
June	1	1 Fry Pan 70 1 Fry Pan 60 4 tin Cups 40	1.70	
		4 Pie Pans 50 1 3 Pt Pan 10	.60	
		3 Pr Knives & Forks 75 4 Spoons 20 2 milk Pans 50	1.45	
		1 Camp Kettle 75 1 coffee Pot 60	1.35	
" "	6	2 Picks	2.25	
		1 Ax Handle	1.55	
		2 Pick Handles	.60	
"	11	2 Shovels 85	1.70	
July	17	1 Pick Handle	.30	
1892 Dec	5	10.# No 1 Powder 22	2.20	
		100 ft Fuse	.80	
		5 # 40d Nail	.30	14.80

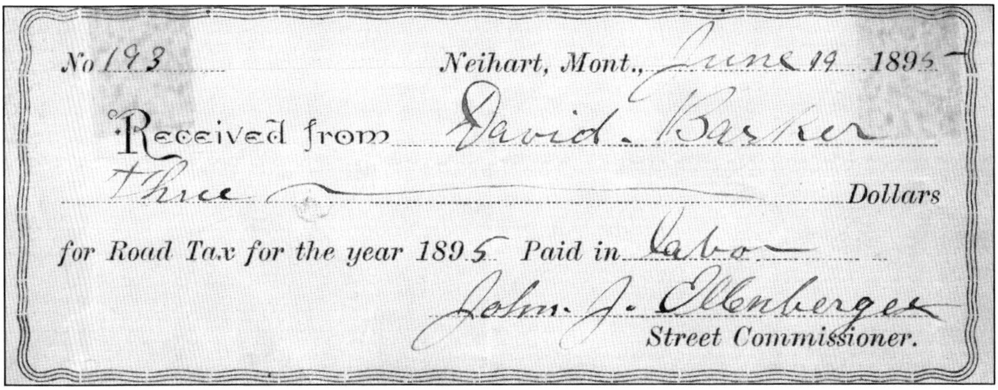

No 193 Neihart, Mont., June 19 1895

Received from David Barker

Three Dollars

for Road Tax for the year 1895 Paid in Labor

John J. Ellenberger
Street Commissioner.

PAYING THE BILLS. The above invoice is for a street fee owed by David Barker to the Town of Neihart in 1895. The invoice below is for freight hauled for David Barker to Lethbridge Coal in January and April. The total was satisfied on April 23, 1896. (Both, D.L.S. Barker Collection, THM.)

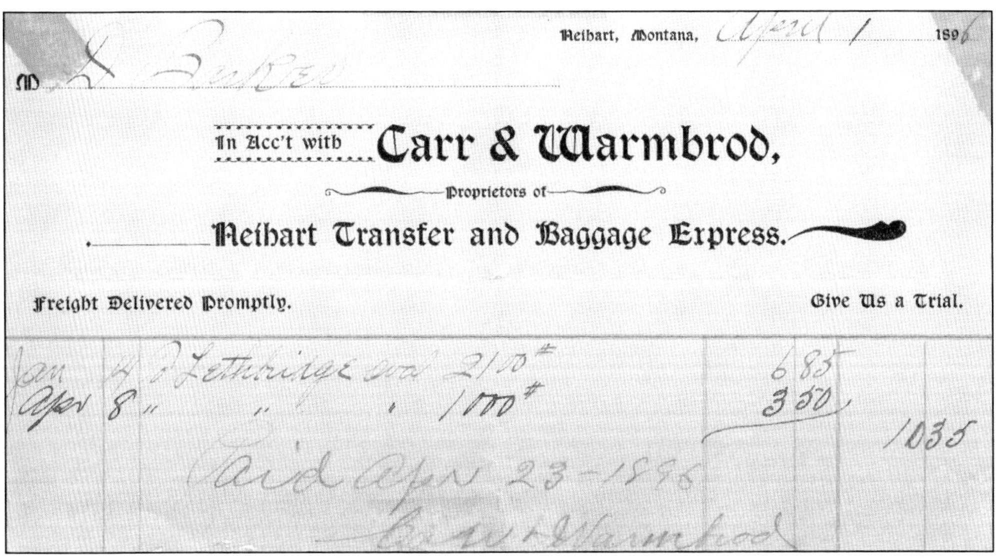

Neihart, Montana, April 1 1896

M D. Barker

In Acc't with Carr & Warmbrod,

Proprietors of

Neihart Transfer and Baggage Express.

Freight Delivered Promptly. Give Us a Trial.

Jan	4	Lethbridge coal 2100#	6 85	
Apr	8	" " 1000#	3 50	10 35

Paid Apr 23-1896
Carr & Warmbrod

Three

Merchants and Business

You may see such . . . places, where only meadows and forests are visible—and will find it hard to believe that there stood at one time a fiercely-flourishing little city, of two thousand souls . . . and now nothing is left of it all but a lifeless, homeless solitude . . . In no other land, in modern times, have towns absolutely died and disappeared, as in old mining regions of California.

—Mark Twain, *Roughing It*

Perhaps half of all the people who populated the early camps were involved with actual mining operations. The rest were there to serve those who did the mining. This second half usually consisted of lumbermen, store owners, hotel and boardinghouse operators, lawyers, ranchers, teamsters, restaurant owners, servers, saloon keepers, perhaps a down-on-his luck physician, and the occasional preacher. Neihart, though, was different; it did not take long for women and families to populate the camp. Schools and churches were an important part of life in early Neihart.

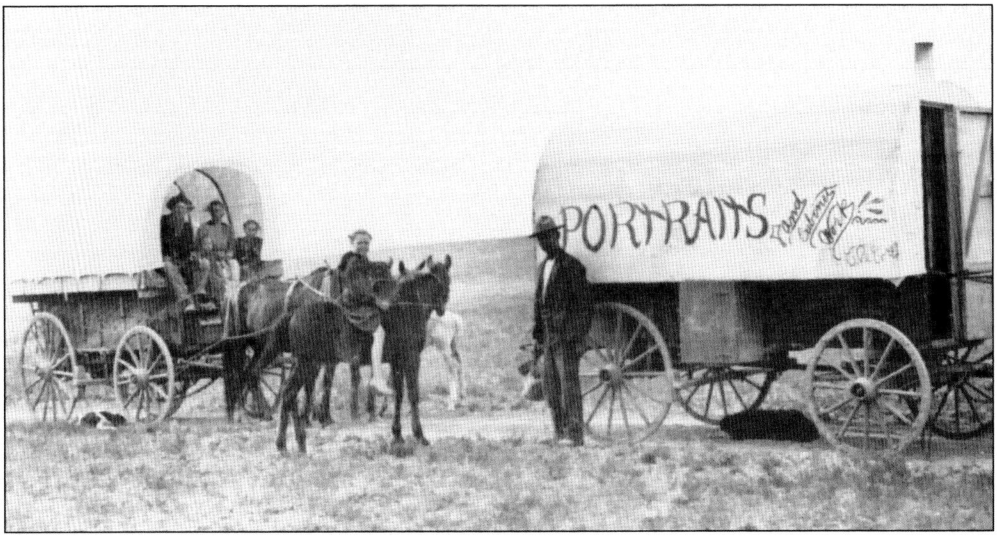

Two Wagons to Transport Itinerant Photographers and Their Families, c. 1890. After the Civil War, photography and cameras improved. Having honed their craft during the war, many itinerant photographers headed west. They followed prospectors in order to mine their version of a bonanza. Photographers made it possible for prospectors to send photographs back to "the states," showing their families that they were doing well and sometimes sending them a picture of their "new digs." Thanks to those photographers, readers are able to see what Neihart looked like in the 1880s and 1890s. (MOR.)

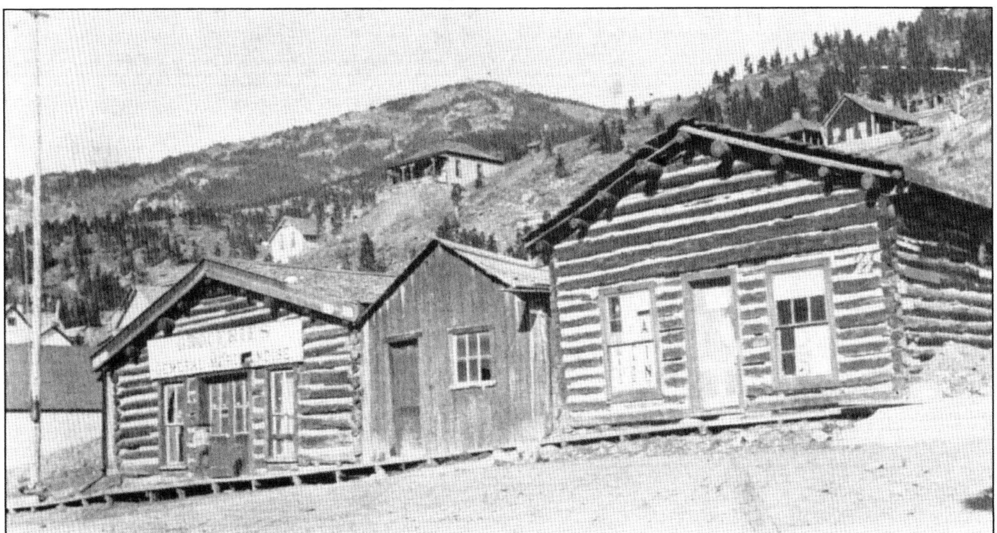

Neihart's First Store. In 1883, Canadian Donald Mackintosh, opened Neihart's first store. It was not much on frills, featuring a dirt floor, poles as shelves, a dirt-covered roof, and a limited supply of goods. Mackintosh robustly encouraged all of his customers to purchase as much as they could handle as quickly as possible. Apparently, his supplier, F.W. Reed of Barker, failed within a few days of the Neihart delivery, and Mackintosh feared the lawyers would claim his merchandise if it remained in his store. The store also served as Neihart's first post office upon Mackintosh's appointment as postmaster on August 21, 1882. He served until September 28, 1895, when Joseph C. Hussey replaced him. (THM.)

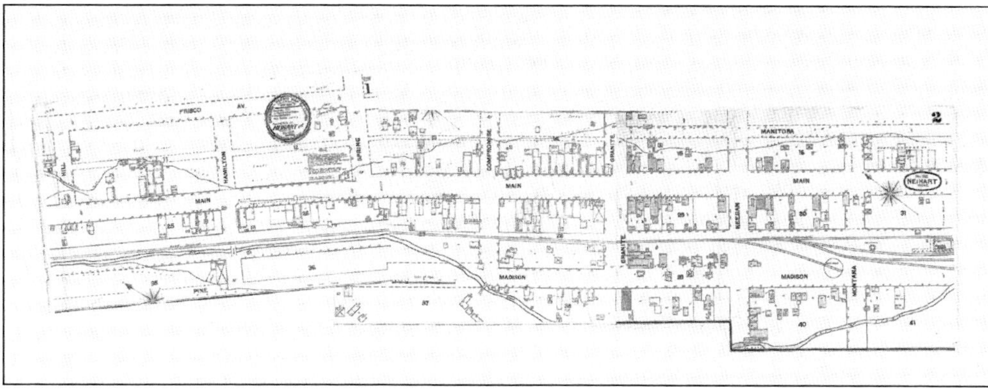

Sanborn Fire Insurance Maps. Sanborn Maps were originally designed for fire-insurance underwriters in the late 19th and early 20th centuries. This Sanborn Fire Insurance Map of 1892 was produced in color since the company inspectors used colors to denote various risks for the underwriters at the home office. At that time, fire-insurance rates were based on the type and quality of construction, as well as the availability of fire suppression or existence of fire departments. The original users of these maps, traveling to their insurance offices via horse-and-buggy or streetcar, would doubtless marvel at the modern technology now used to rate fire insurance risks. (MHS.)

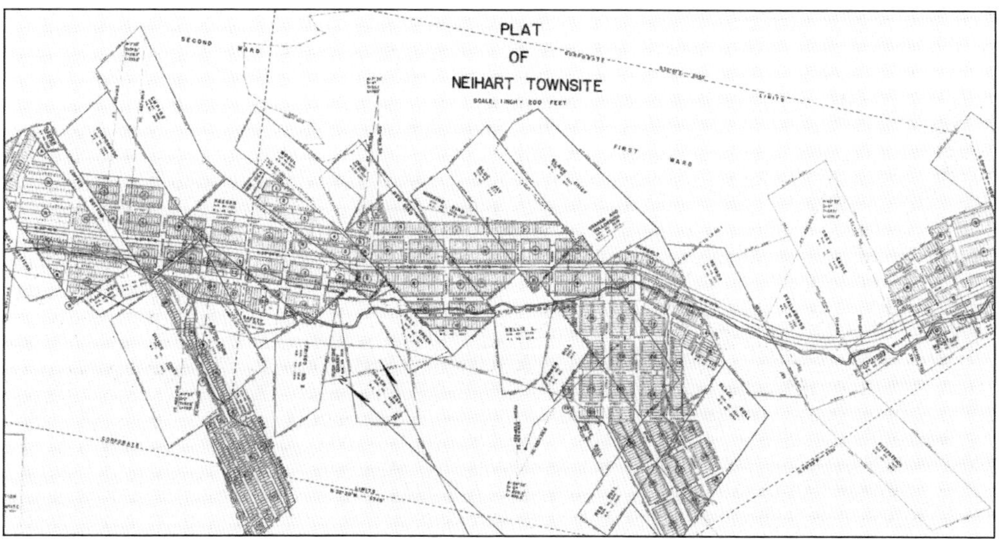

The Plat of a Promising Neihart Townsite. Both the date of this plat and the person responsible for its commission are unidentified. The plat, however, is interesting for its grandeur and speculation of the potential for Neihart and the value of the ore-bearing lodes. The section extending down in the middle third of the image represents the O'Brien Creek area. Had that growth actually taken place, the concentration of homes along O'Brien's watershed would have challenged that of San Francisco, with dwellings atop dwellings, as the streets climbed ever higher on the slopes of the mountain. (Author's collection.)

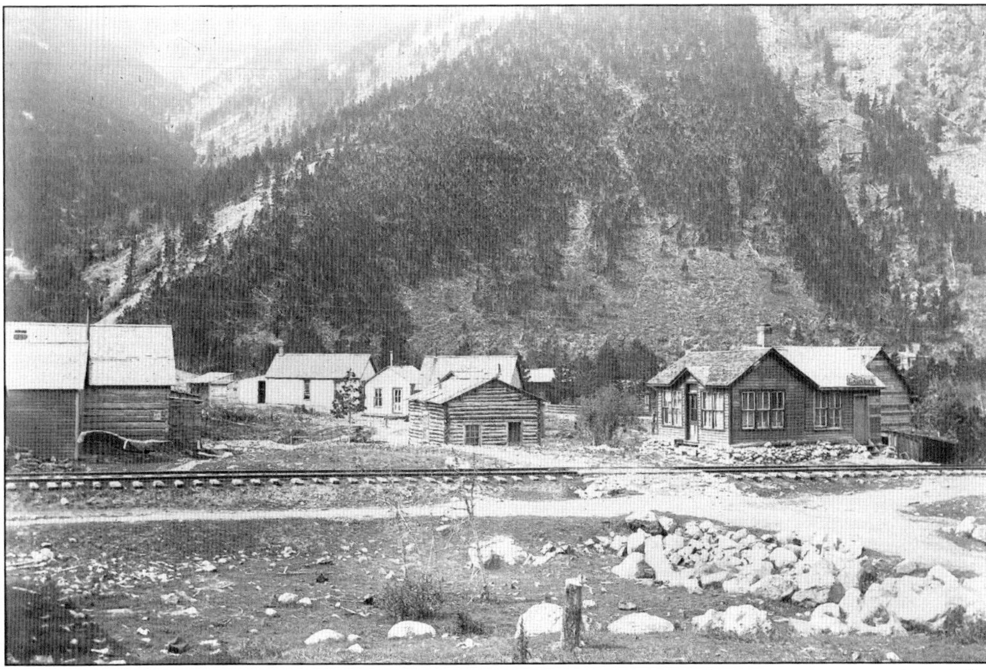

COMPROMISE STREET WEST TOWARDS BELT CREEK AND JOHNSTON GULCH, c. 1893–1894. The large log building on the right has a sign proclaiming either City Laundry or US Laundry. On the 1892 Sanborn Fire Map of Neihart is a reference to a log laundry by the tracks at the end of Compromise Street leading to Belt Creek. The *Neihart Gazetteer* for 1892 shows no City Laundry, but there is a US Laundry. According to the 1892 *Gazetteer*, the proprietor of the US Laundry was J.P. Ryan, with Carl Orstrum and Annie Kirkendall as his colleagues. (MHS.)

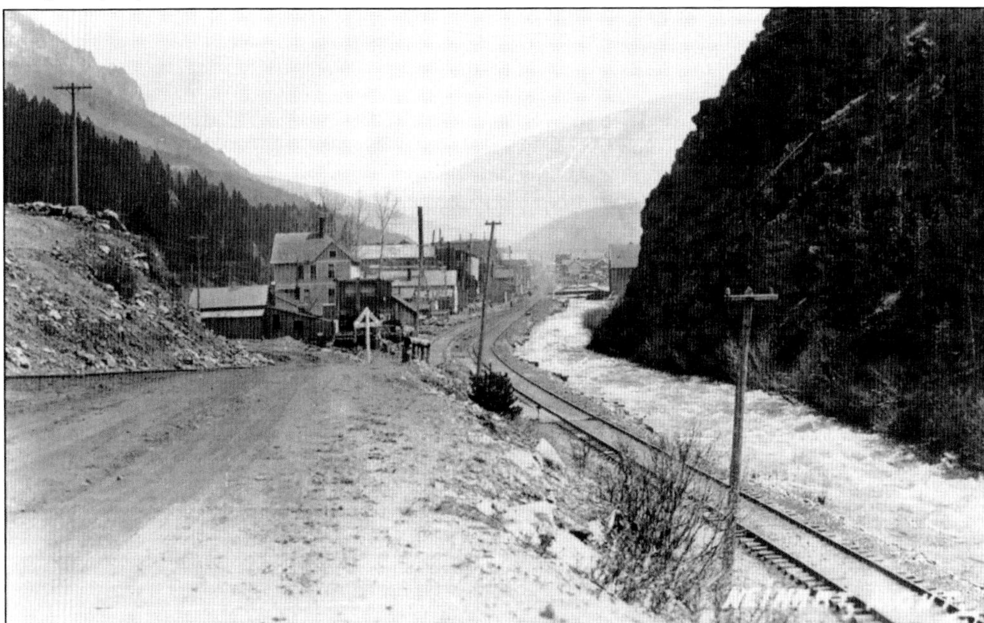

ENTERING NEXT TO THE QUEEN OF THE HILLS. This photograph provides a wider view of the entrance into Neihart as one passes the Queen of the Hills Mine on the left. (Carol Zehntner.)

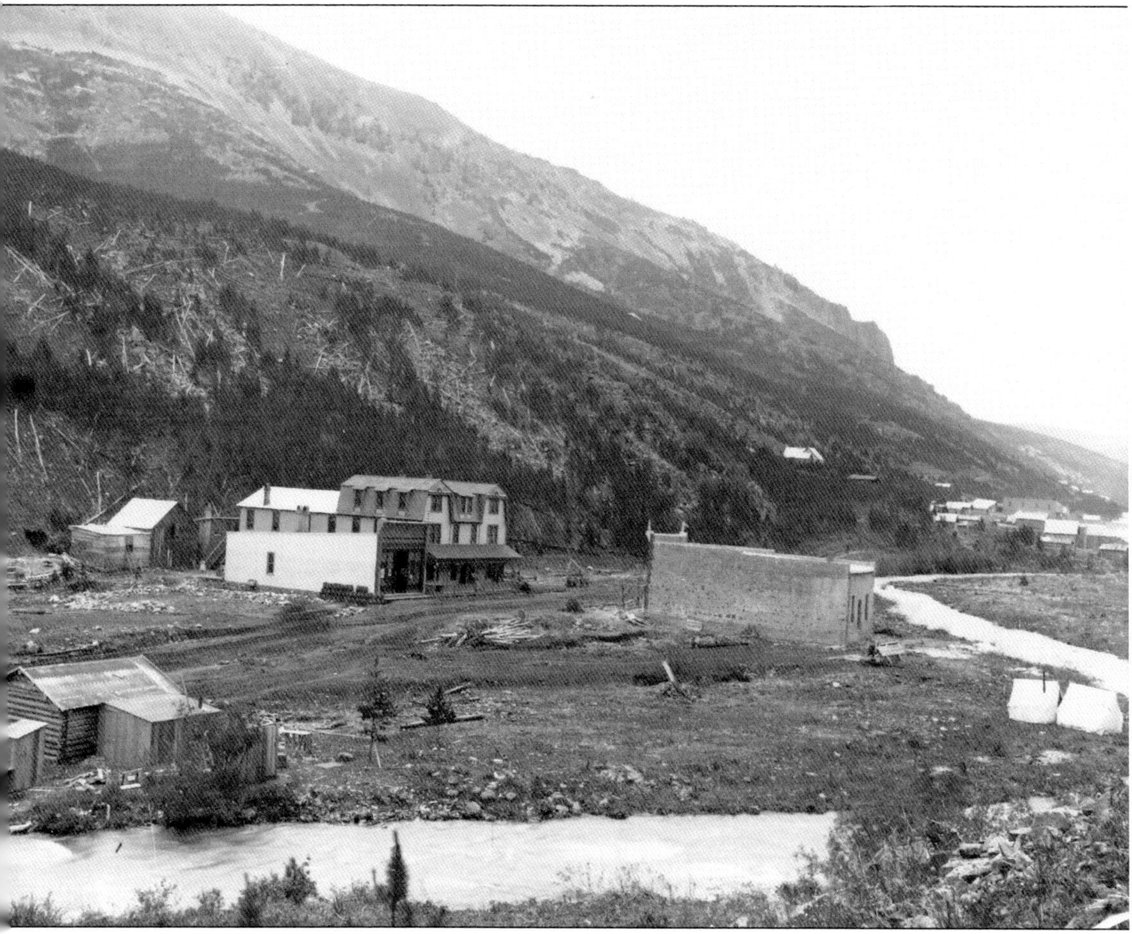

The Pain was in Tents, Until the Frisco Opened. A booming mining camp was not a place of quiet and calm. The camp would have been a beehive of activity 24 hours a day in the beginning. While it is known that Neihart was built of brick and board after 1890, prior to that it should not surprise anybody that a large number of prospectors took shelter in walled tents. Some tents even remained after the railroad arrived, but the majority of miners either built cabins or stayed in boardinghouses. But in 1890, the finest lodging in the Montana Territory was ready and open for business for any prospector, miner or visitor. This image of the Frisco Hotel sharing space with an unnamed saloon operation provides a wonderful view of the north end of Neihart before it burst with growth. (THM.)

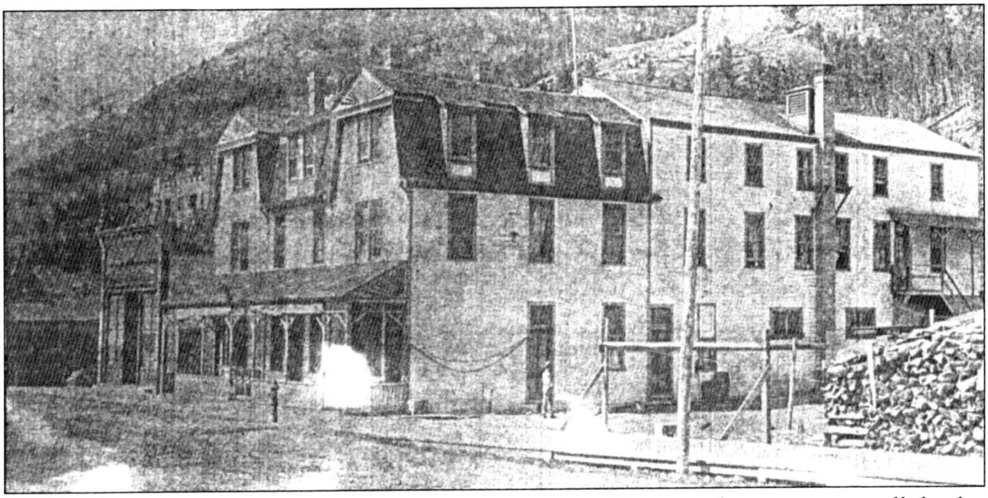

Frisco Hotel Ready to Open. Many lived in or conducted business from tents or small shacks. After all, as most prospectors, miners, and laborers wrote, the only things they needed were a decent bed, a roof over their head, and a good meal. If they had wandered over to the Frisco Hotel, they could have met all three needs in one plush spot. Next door, to the left of the Frisco, was an unidentified saloon operation. (THM.)

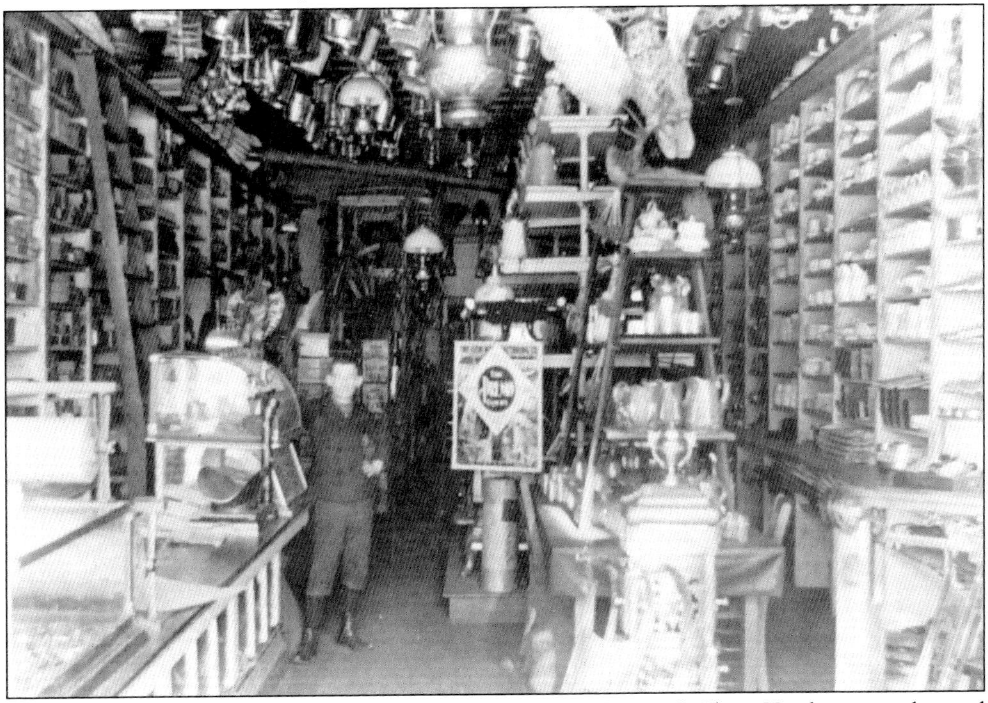

Interior of Lowry & Shaw Hardware Store, c. 1897. Lowry & Shaw Hardware was located on the end of the Pierse Block closest to Granite Street, facing Main Street and Johnston Gulch. Lowry & Shaw carried a full line of shelving and heavy hardware for mining purposes, miners' supplies, stoves, tinware, crockery, lamps, and builders' hardware. Ernest A. Shaw, in addition to his efforts to provide miners with the best merchandise available, also served the community as mayor for one year, in 1894. The Shaw family lived above the store. Ed Bennett, the son of town attorney Richard Bennett, is pictured here. (THM.)

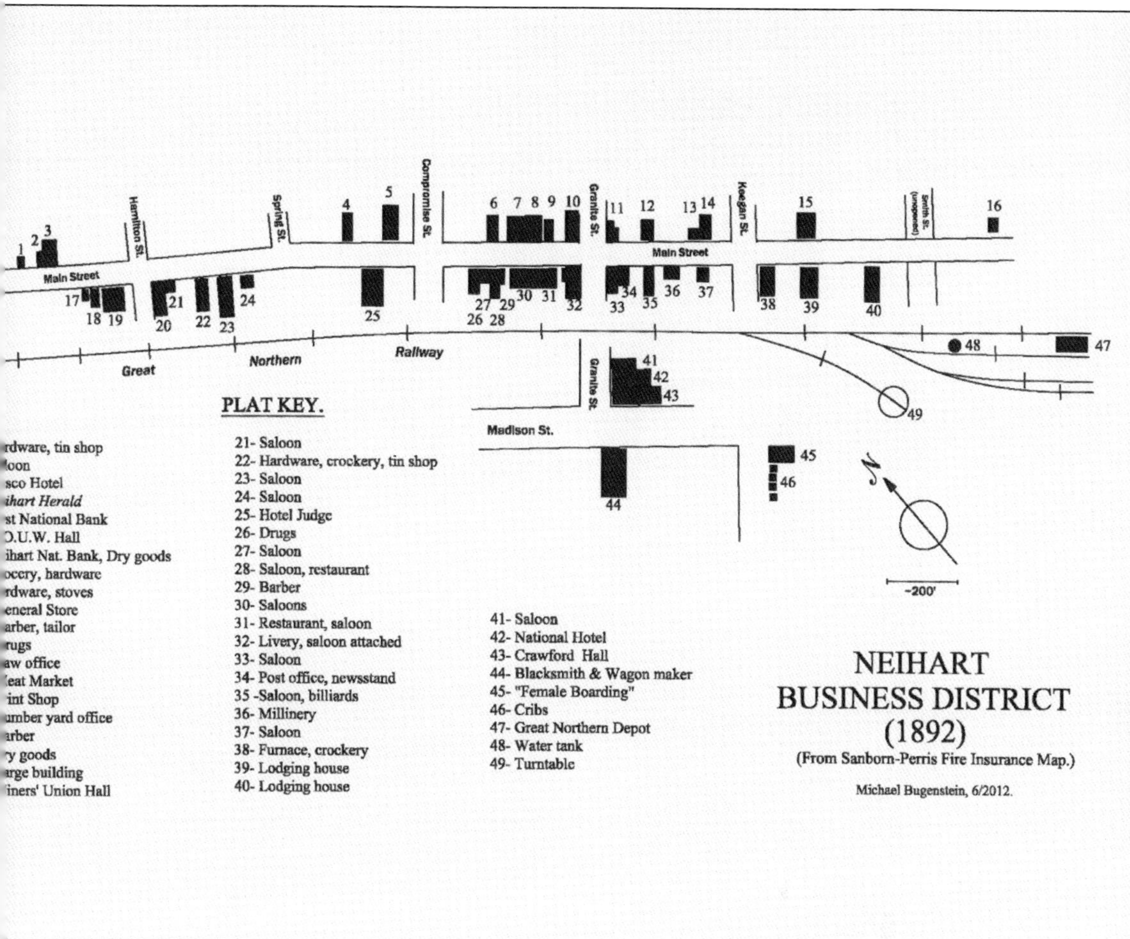

Business District. Based on the 1892 Sanborn Fire Insurance Map, this plat reveals the locations and names of several Main Street businesses. Bob's Place (motel, dining, gas, and bar) serves as a current landmark, located on the west side of Main Street between Compromise and Granite Streets occupying almost an entire block and extending approximately from No. 26 (drugs) to No. 31 (restaurant/saloon). (Author's collection.)

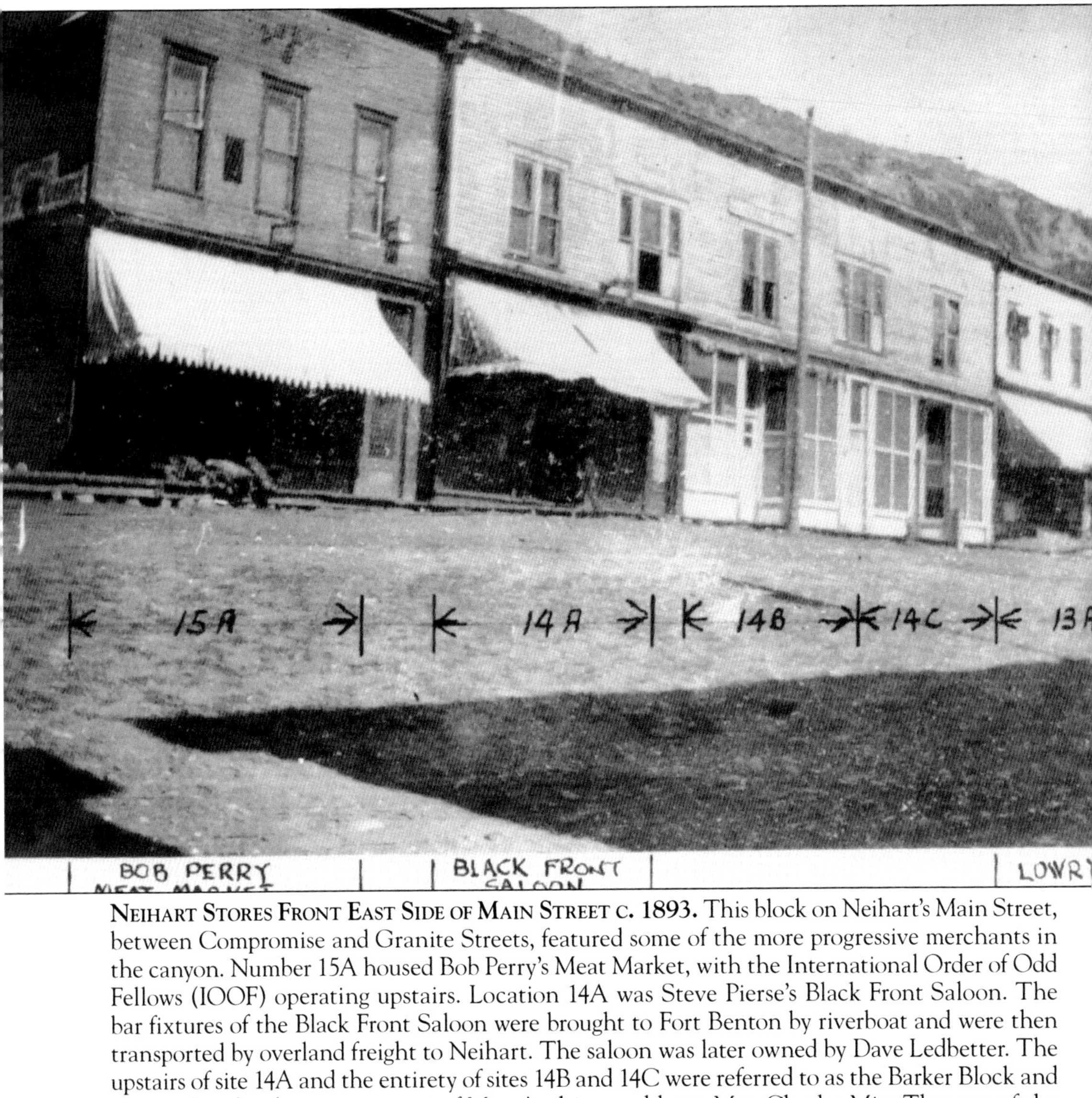

NEIHART STORES FRONT EAST SIDE OF MAIN STREET C. 1893. This block on Neihart's Main Street, between Compromise and Granite Streets, featured some of the more progressive merchants in the canyon. Number 15A housed Bob Perry's Meat Market, with the International Order of Odd Fellows (IOOF) operating upstairs. Location 14A was Steve Pierse's Black Front Saloon. The bar fixtures of the Black Front Saloon were brought to Fort Benton by riverboat and were then transported by overland freight to Neihart. The saloon was later owned by Dave Ledbetter. The upstairs of site 14A and the entirety of sites 14B and 14C were referred to as the Barker Block and operated under the management of Mrs. Andries, and later, Mrs. Charles Mix. The uses of the block are not known. The main floor of site 13A housed two businesses: Lowry & Shaw Hardware Store occupied the right side of the building, and Pierse & Harrison Grocery used the left. The upper floor of 13A was the living quarters of the Ernest A. Shaw family. (THM.)

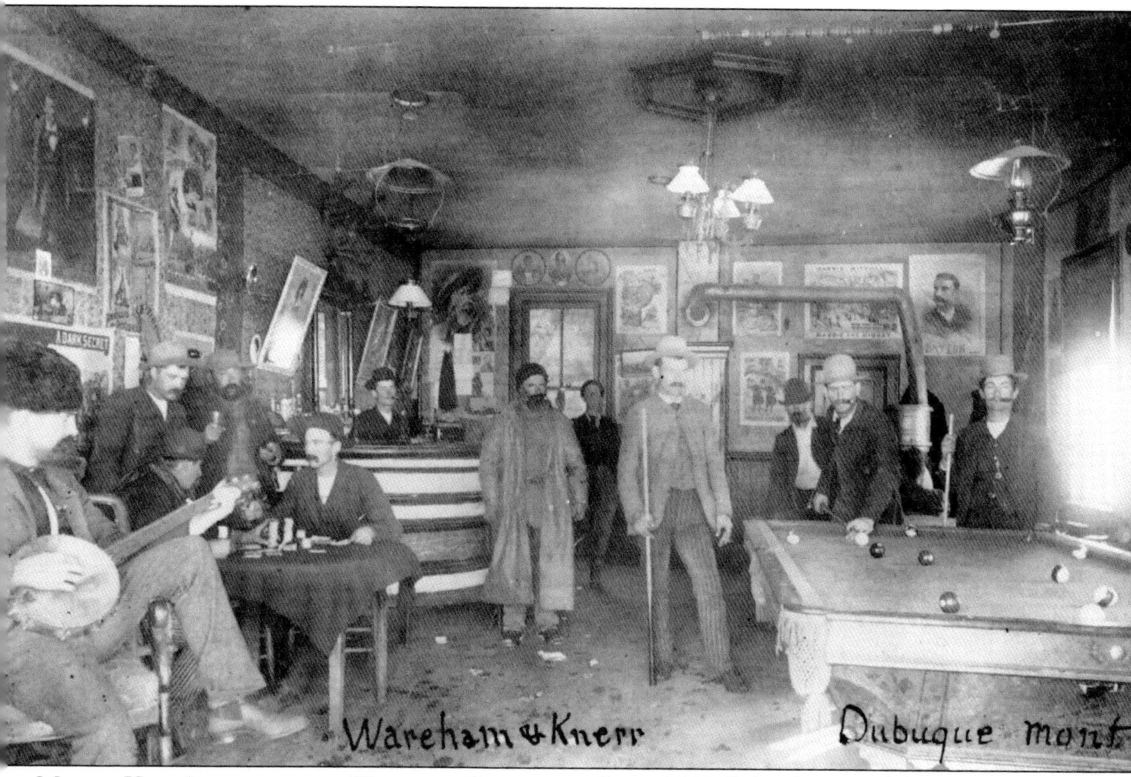

MINERS KEPT LUBRICATED BY NINETEEN SALOONS, C. LATE 1892. It has been said that Neihart had 18 saloons. Maybe. The 1892 Sanborn Fire Map does confirm 15, but adding all three of the hotels; the White House, which was a saloon; as well any tents or shanties missed in the count, the number rises to at least 22. Neihart was no different than any other mining camp of the far West, with saloon owners often setting up shop in a tent (with two kegs supporting rough-cut boards to serve as the bar) before committing to a building. Strangely, no photographs survive of Neihart's saloons, but this photograph from nearby Dubuque, Montana, offers an inside view into a similar establishment. Although no images exist, Neihart's saloons are remembered in print. A November 5, 1932 article in the *Great Falls Tribute* references a "famous resort" in the town, and a spot called The Combination was apparently a rather popular gathering spot for several years. Ralph Bemis wrote this in the *Tribune* about another spot called Mix & Johnston's: "Every brand of humanity which obeyed the call to mining camp life has passed its portal at one time or another . . . enact[ing] high revelry, comedy, tragedy, and despair." A better summary has not been written. (MHS.)

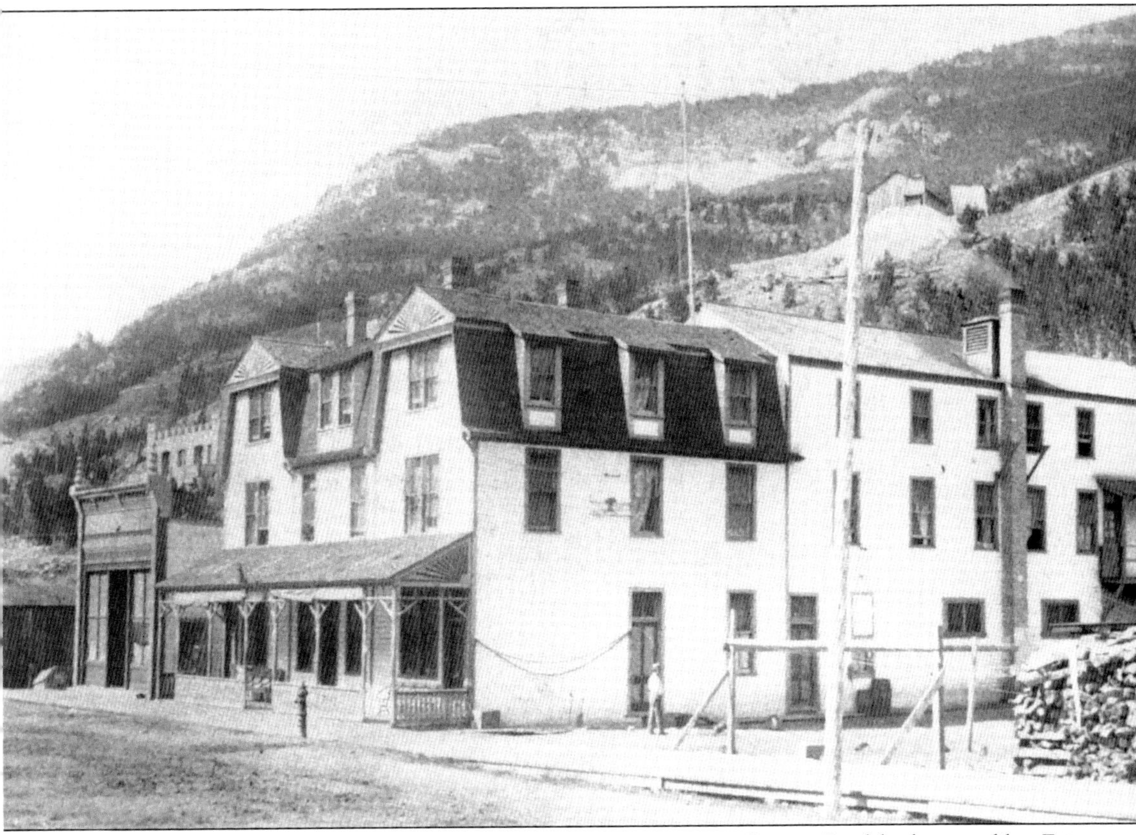

ONLY FIRST-CLASS HOUSE IN THE CITY, C. 1890. Every time George Roehl advertised his Frisco Hotel, he used the above description, and first class it was. Opened on October 1, 1890, and twice expanded, the Frisco was elegantly furnished, offering 60 airy rooms (all with electric lights) and water on all three floors. Roehl, originally of Stralsund, Prussia, immigrated to the United States in 1870 with his wife, Amelia (née Rampe), a native of the German-possessed island of Rügen in the Baltic Sea. First settling in Allentown, Pennsylvania, they then lived in Minneapolis, Minnesota, before migrating to Montana Territory. Prior to the Frisco, George and Amelia oversaw the food, lodging, and service in Great Falls's Milwaukee House for two years. The Roehls had 10 children, two of whom, Paul and Charles, remained in Neihart as miners. Paul, however, felt the spirit of public service and was elected to the Neihart Council as an alderman around 1910. (THM.)

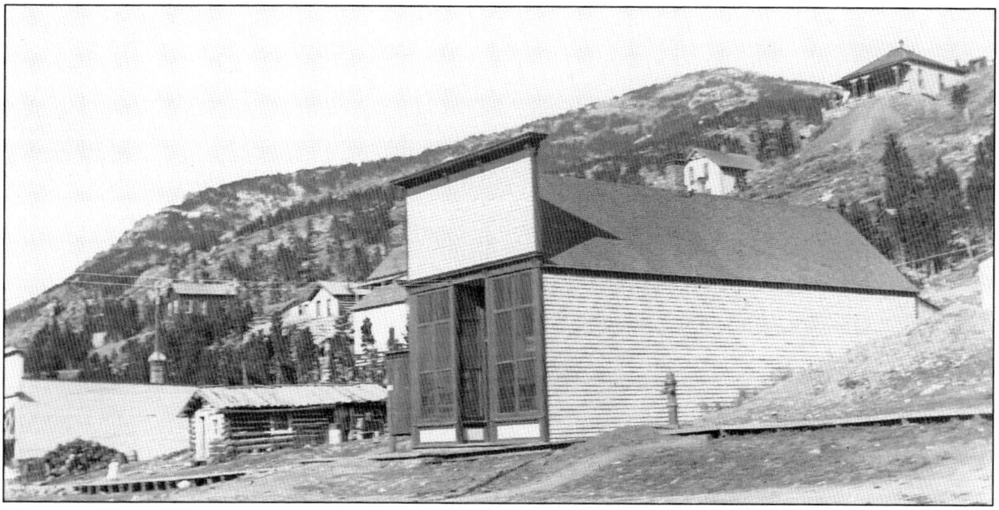

NEIHART BANK NEIGHBORHOOD ONE YEAR APART. The above photograph has been identified and archived as the Neihart Bank. It is agreed that the building is one of Neihart's banks. Looking to the far left, past the log cabin and the pile of logs, is the building housing the *Neihart Herald* newspaper. All of this is located between Spring and Compromise Streets on Main Street. Notice the unidentified girl on the boardwalk in front of the log structure. Across Main Street would be the Hotel Judge. The Sanborn Fire Insurance map, dated 1892, has this structure labeled as 1st National Bank, with the Neihart Bank being a block farther down between Compromise and Granite Streets, both on the same side of Main Street. This photograph was taken in 1891. The photograph below, taken a year later, features a fancy horse buggy and five unidentified men posing in the middle of Main Street, at the intersection of Compromise and Main Streets. Four of the five have cigars clenched in their mouths with one (in front of the white canvas cover on wagon) perhaps wearing sunglasses. The important fact is that Neihart was enjoying great growth, prosperity, and fortune—enough to require more than one financial institution. However, the silver crash of 1893 is lurking. (Both, MHS.)

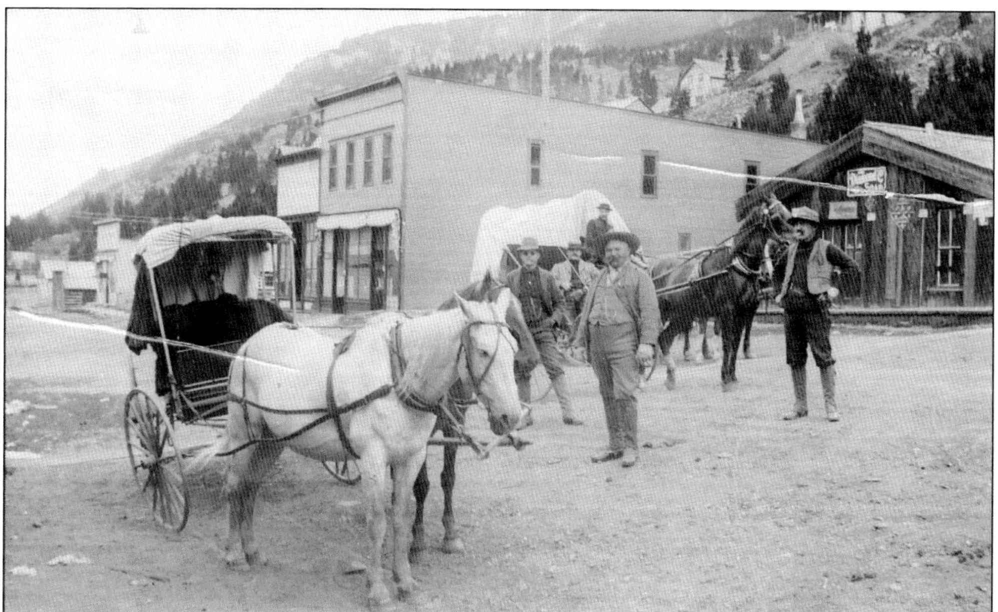

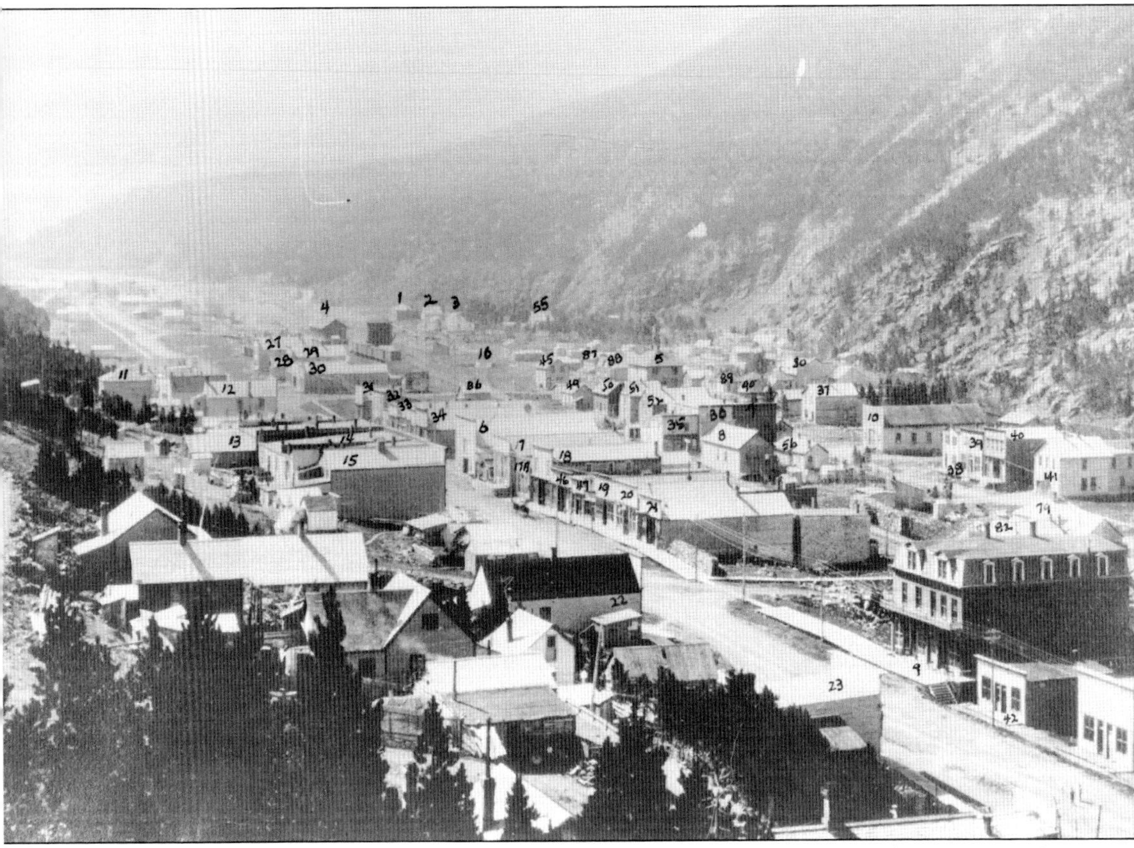

NEIHART FROM ROCK CREEK LOOKING SOUTH, 1893. This view, from the summer of 1893, shows Neihart's central business district. Dick Bennett, a member of the pioneer Bennett family and resident of Great Falls, has provided a key to the identity of many of these businesses and structures. Some highlights include the Carroll Bennett house, now the Charlie Swanson home, still standing (No. 1); Judge Hotel (No. 9); Murphy's Saloon (No. 19); the fire department, located on the second floor of city hall (No. 32); the school (No. 73), and the Richard Bennett law office (No. 82). (Dick Bennett.)

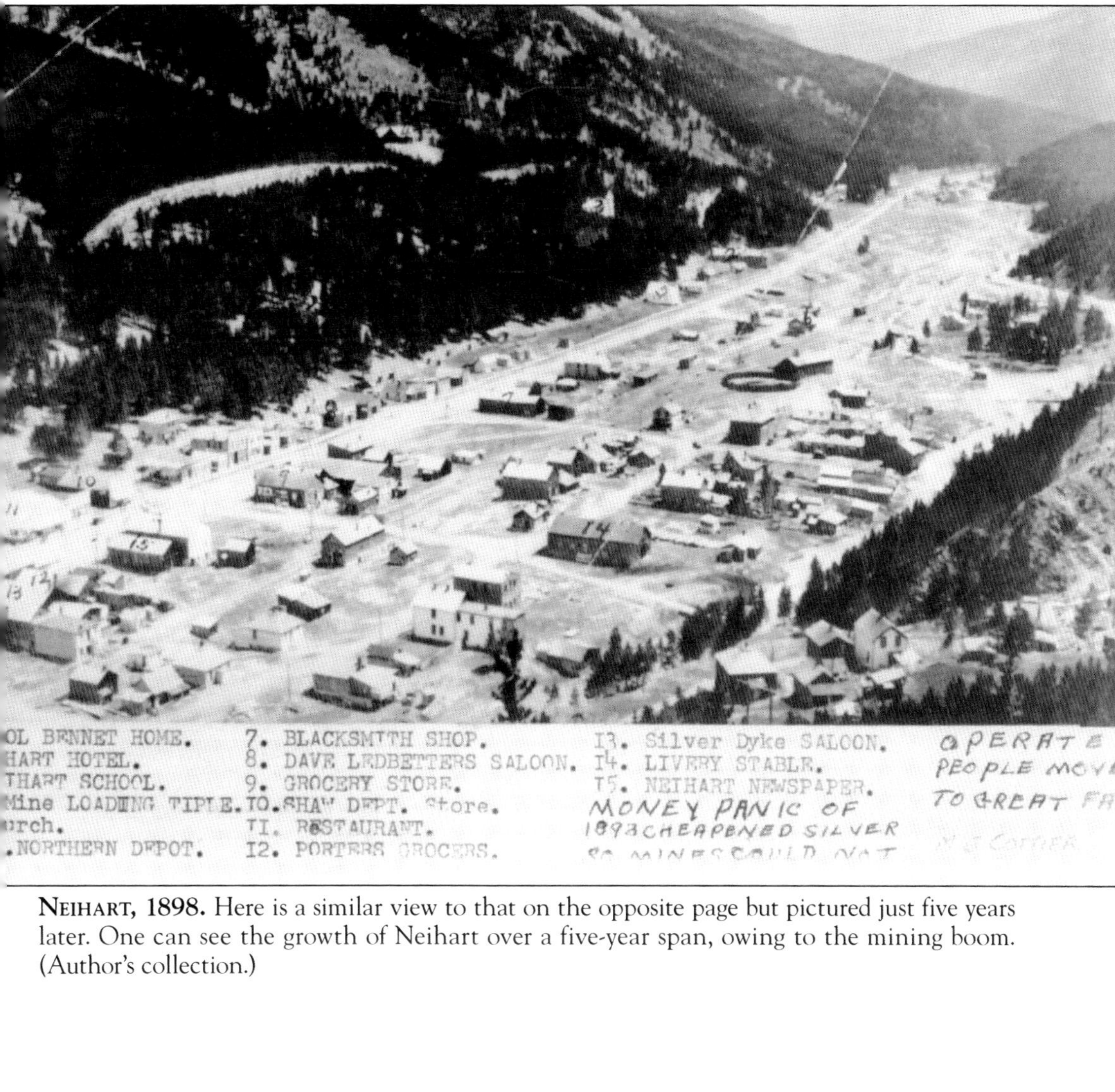

NEIHART, 1898. Here is a similar view to that on the opposite page but pictured just five years later. One can see the growth of Neihart over a five-year span, owing to the mining boom. (Author's collection.)

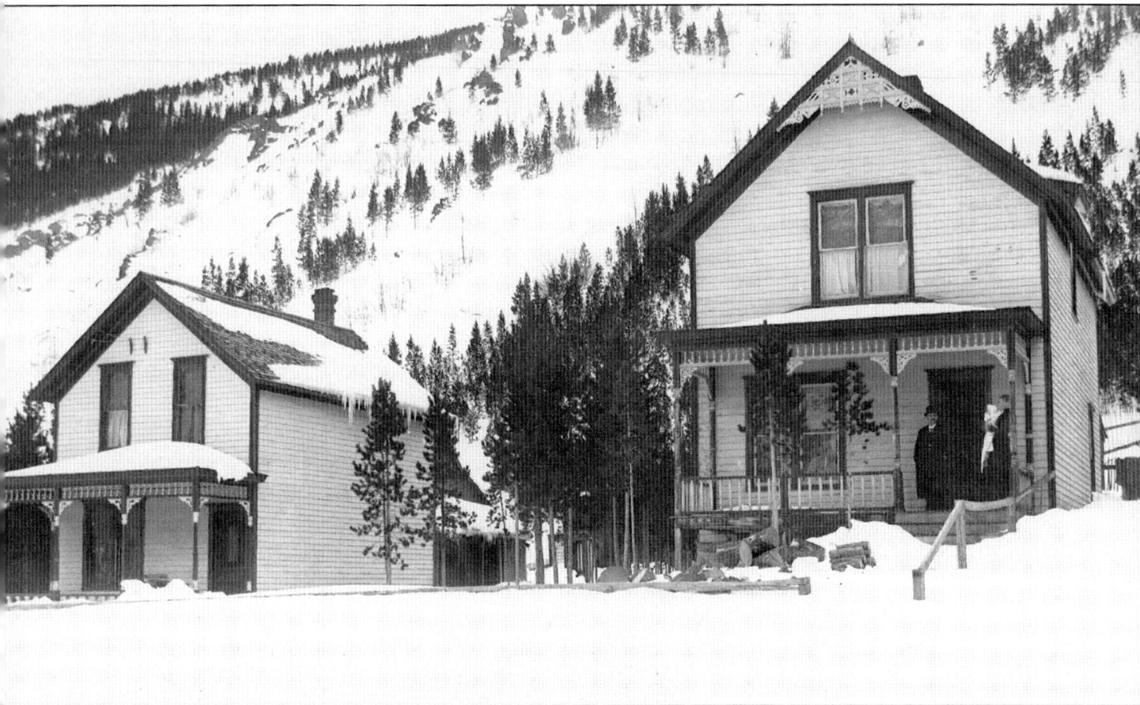

Harrison Home, South Neihart. The house on the right belonged to William H. and Sarah C. Harrison. Their daughter Alice was a close friend of Myrtle Neihart, and the two enjoyed summer romps and picnics together. Though unidentified, it is safe to assume that William and Sarah are posing, with a baby, on the right side of the front porch. The Harrisons, according to the 1900 census, had three sons (Frank, Alexander, and William P.) and one daughter, Alice. William was a partner with Steve Pierse and managed their grocery store in Neihart. The store was located in the Pierse Block on Main Street between Compromise and Granite Streets. Pierse & Harrison Grocery was located across Main Street from current day Bob's Place. A Californian by birth, Harrison moved to Nevada with his parents, then to Utah, and, finally, to Montana. Besides merchandising, he served in the Montana Legislature, was assessor for Cascade County, and then city clerk for Great Falls. Both houses in the photograph are occupied today. (MHS.)

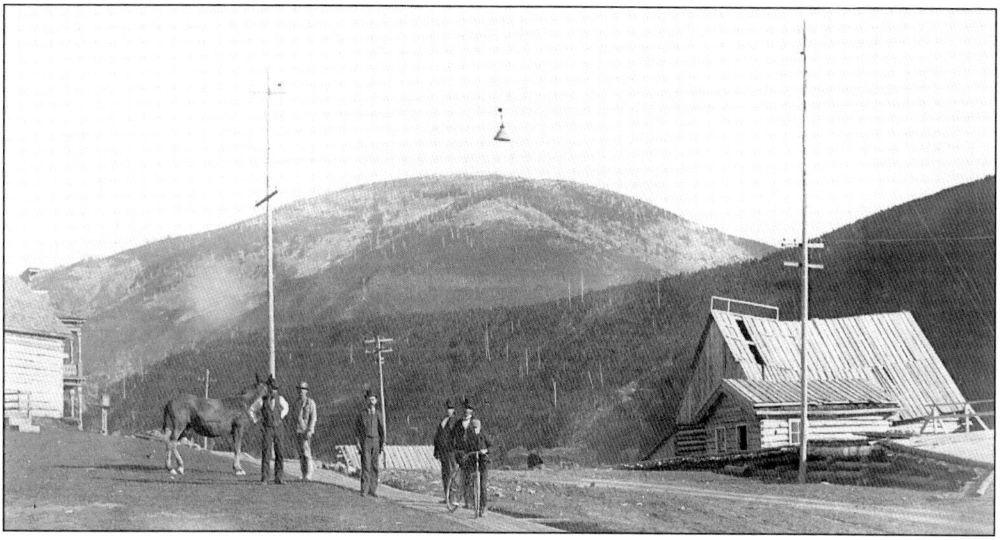

NEIHART'S FIRST STREETLIGHT, C. 1892. Neihart's residents took great pride in their first streetlight, and the Queen of the Hills Mine became the first operation in Montana to try electrical machinery in mining. As more area conveniences became powered by electricity, George W. Brown (cashier of the Bank of Neihart) and B.F. Milan (of Sioux Falls, South Dakota) made application to the city in January 1891 for a franchise to operate an electric street railway to be up and running within six months. Nothing ever became of that proposal. (MHS.)

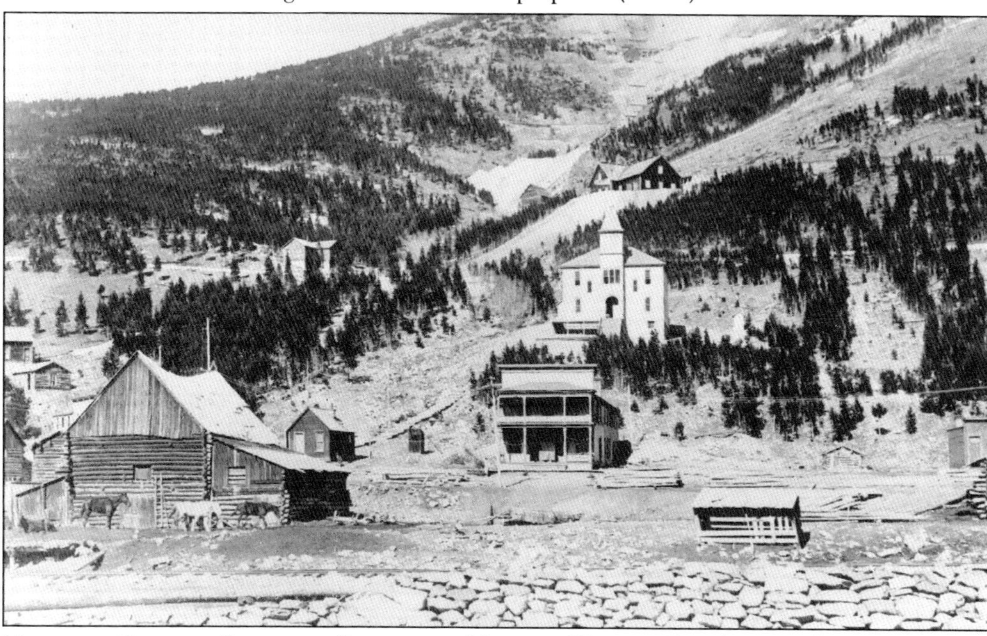

MONTANA CENTRAL BARN AND STABLE AND NEIHART HOTEL. This photograph, taken from the rail yard looking east-northeast up the Broadwater Mining claims, shows the tailings (waste) taken from the mines. These tailings eventually led to state and federal warnings about toxic substances flowing into Belt Creek. In front of the school is the Neihart Hotel, with Bill Spence's home to the left. Behind the barn is the upper back corner of the miner's boardinghouse, an unidentified log structure, and the Fred Hart house above them all. Just to the left of the barn is the corner of the C.H. Burchard home, near Smith Street. (MHS.)

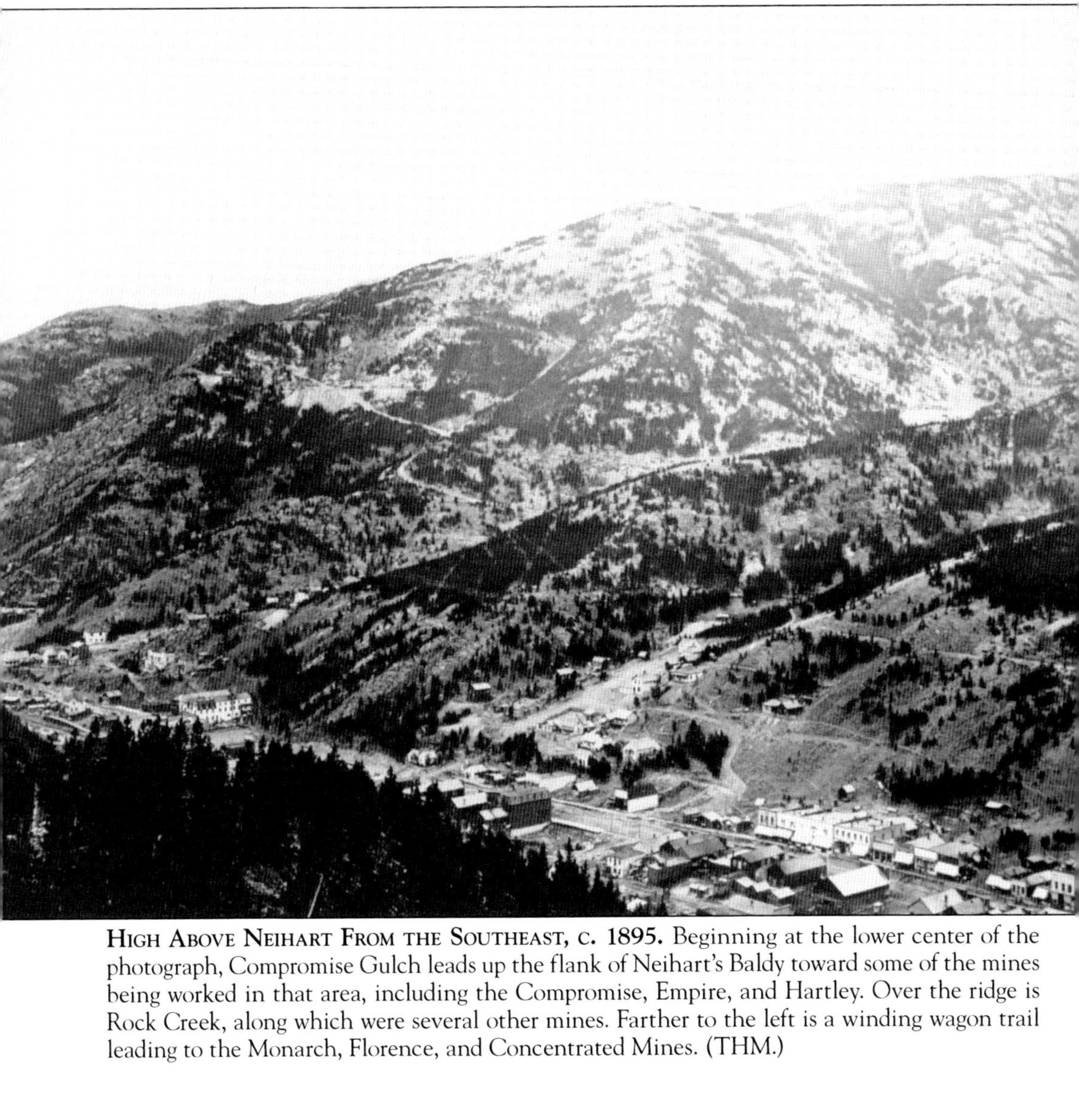

High Above Neihart From the Southeast, c. 1895. Beginning at the lower center of the photograph, Compromise Gulch leads up the flank of Neihart's Baldy toward some of the mines being worked in that area, including the Compromise, Empire, and Hartley. Over the ridge is Rock Creek, along which were several other mines. Farther to the left is a winding wagon trail leading to the Monarch, Florence, and Concentrated Mines. (THM.)

Dr. C.E.K. Vidal's Arrival, 1893. Vidal, born on February 6, 1870, in Sussex, England, practiced medicine in Neihart and Belt from 1890 to 1895. Dr. Vidal married Mable Macrae, of Montreal, in 1894. He was active in the Neihart community as a member of and medical examiner for the Belt Mountain Lodge No. 18 of the Ancient Order of United Workmen (AOUW), as well as for New York Life Insurance and other insurance companies; he also owned a half interest in the Belt Mountain Hospital in Neihart. The AOUW was instrumental in forming the Belt Mountain Hospital, collecting dues from members and paying sick or death benefits. It may have the distinction of being the first Health Maintenance Organization (HMO) in the United States. (THM.)

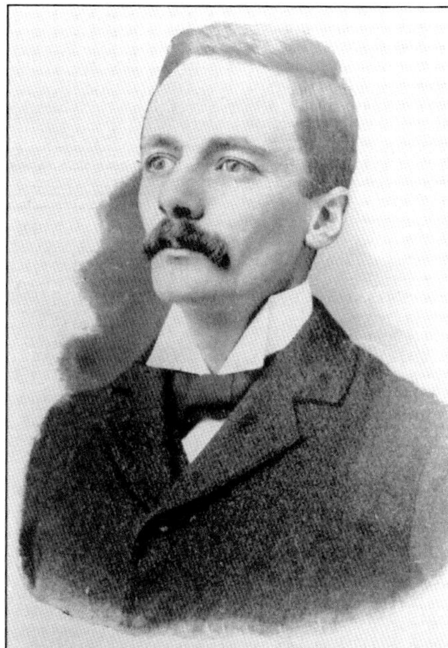

DR. C. E. K. VIDAL.

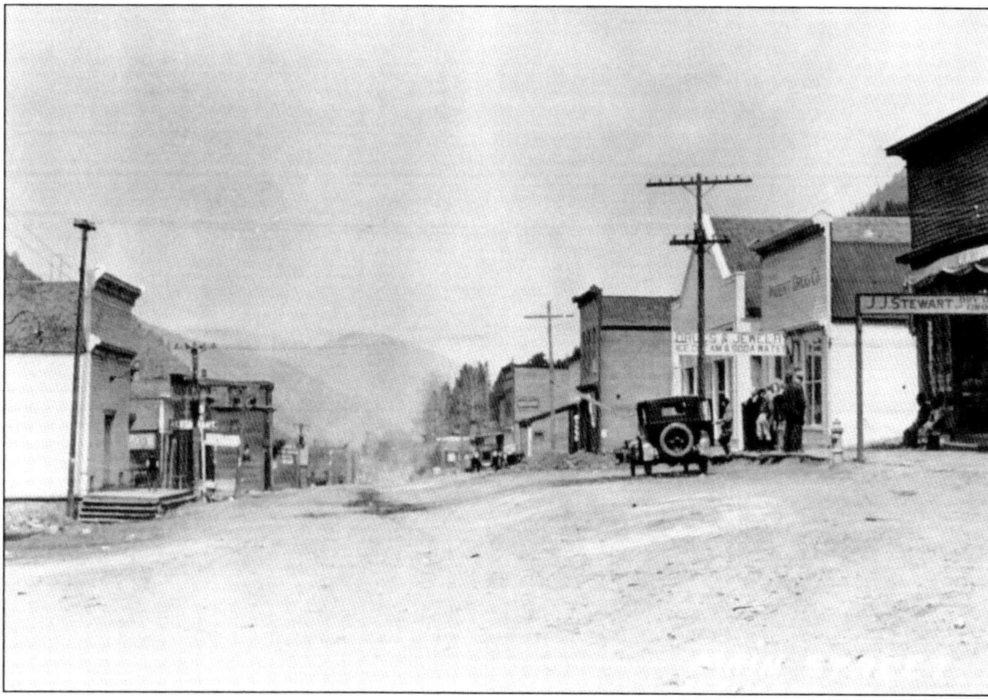

Neihart Main Street, Dusty and Nearly Deserted. This photograph of downtown Neihart appears to have been taken around the time of Prohibition. Many residents of the area participated in the manufacture and running of illegal hooch, just as any self-respecting, mountain-living, hardrock miner would be expected to do. The story goes that the runners had a secreted trail (perhaps Oti Park trail?) at the end of Hoover Creek, which would drop them out at Barker. This image shows what appears to be a dust devil smack-dab in the middle of Main, in front of J.J. Stewart's Dry Goods store and the Drug & Jewelry store next door. (Carol Zehntner.)

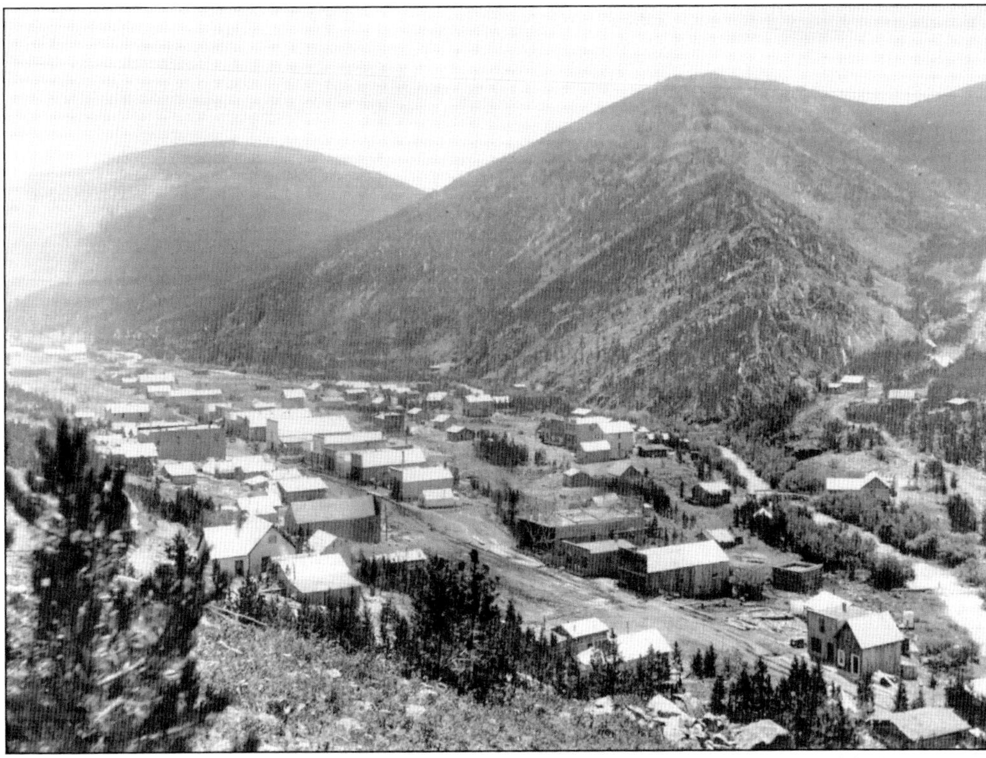

The Beginning of Neihart's Building Frenzy, July 25, 1891. These wonderful photographs show projects, both completed and under construction, in anticipation of the long-awaited and absolutely required transportation service. The railroad ensured prosperity and growth for both the mines and the larger community. This view from atop Compromise Gulch overlooks Forest Street (running behind the structures on Main Street), as well as Main Street and Madison (closest to Belt Creek). On the far right, midway up the photograph, Johnston Street can be seen running up Johnston Gulch. The building being constructed in the center of the image is the Judge Hotel; it would be completed in early September 1891. (Both, MHS.)

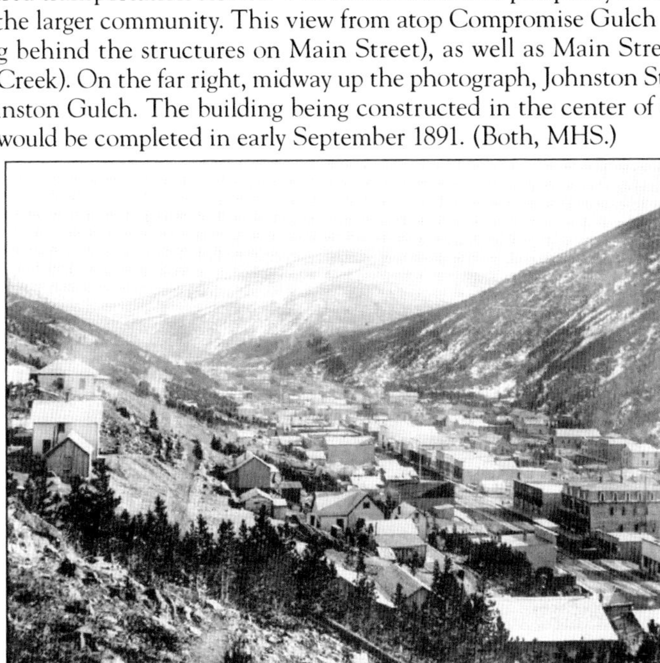

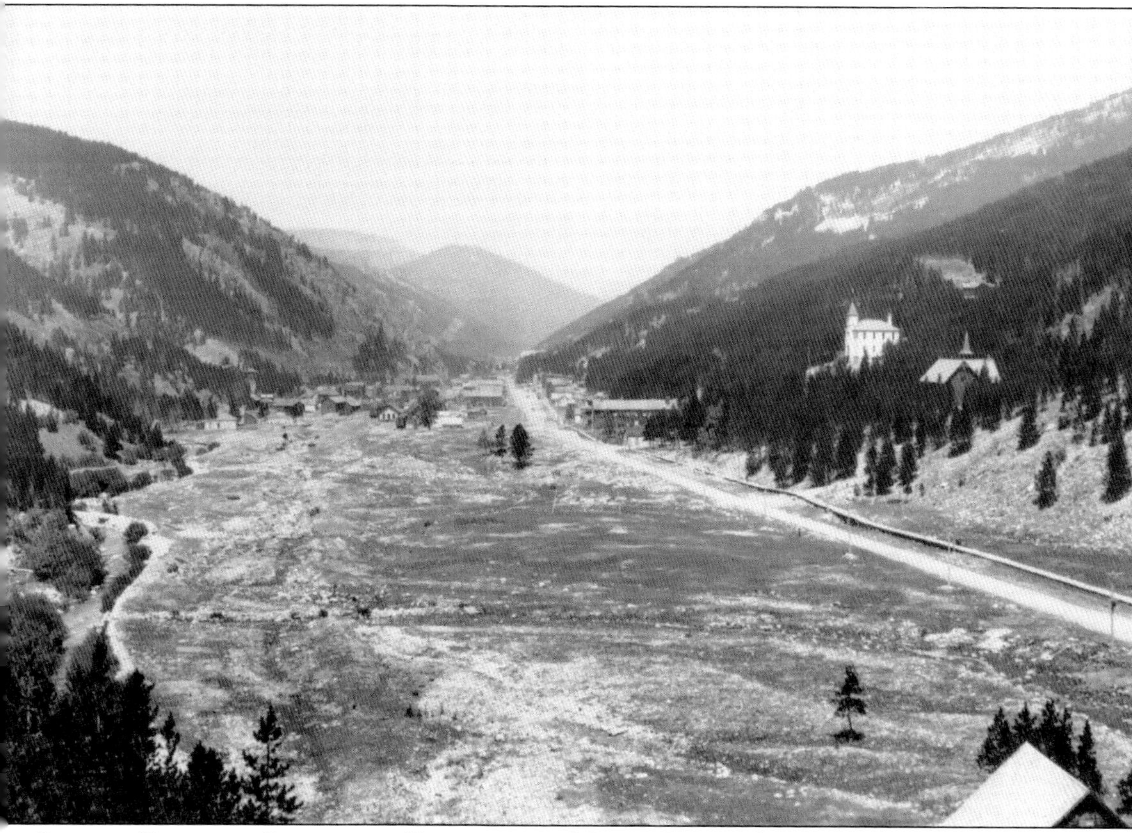

Looking Down the Canyon into Neihart from O'Brien Creek. This photograph shows some interesting features of Neihart. The white line that runs parallel to Main Street was a wooden boardwalk, constructed the entire length of the town in order to keep boots clean and dry. What appear to be telephone poles stand alongside the street, or perhaps they are power or telegraph poles to White Sulphur Springs. To the right in the trees, the steeple of the Methodist Episcopal Church (corner of Forest and Hulbert Streets) is visible, near the white school. Just below the school is the Neihart Hotel. (THM.)

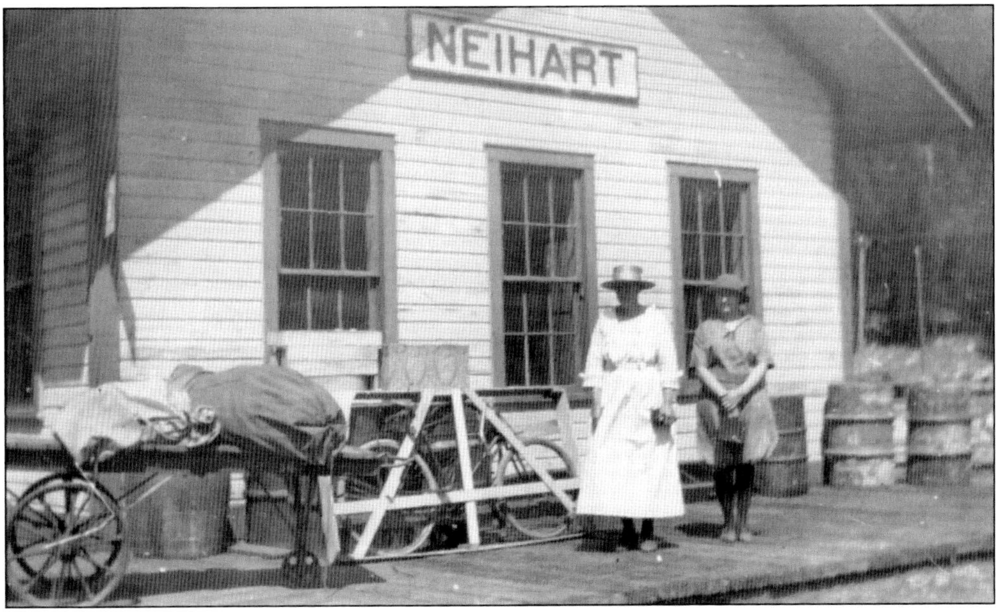

DOWNTOWN SCENES IN EARLY 1900S. The above photograph shows Leona (left) and Esther Carter standing on the platform in front of the Montana Central Passenger Depot in 1919. This building is still in use in downtown Neihart today. After the structure's successful run as the depot, it was occupied by the Catholic church, which conducted mass there for several decades. Within the past few years, the structure has been extensively remodeled; it is now a private cabin fronting old Main Street, or US Highway 89. The photograph below is a c. 1919 view down Neihart's Main Street. Notice that the unidentified gentleman walking away from the camera, on the left, has no boardwalk or sidewalk on which to tread, while the unidentified individual across the street appears to have more solid footing. (Above, MHS; below, Mary Sohl Collection, THM.)

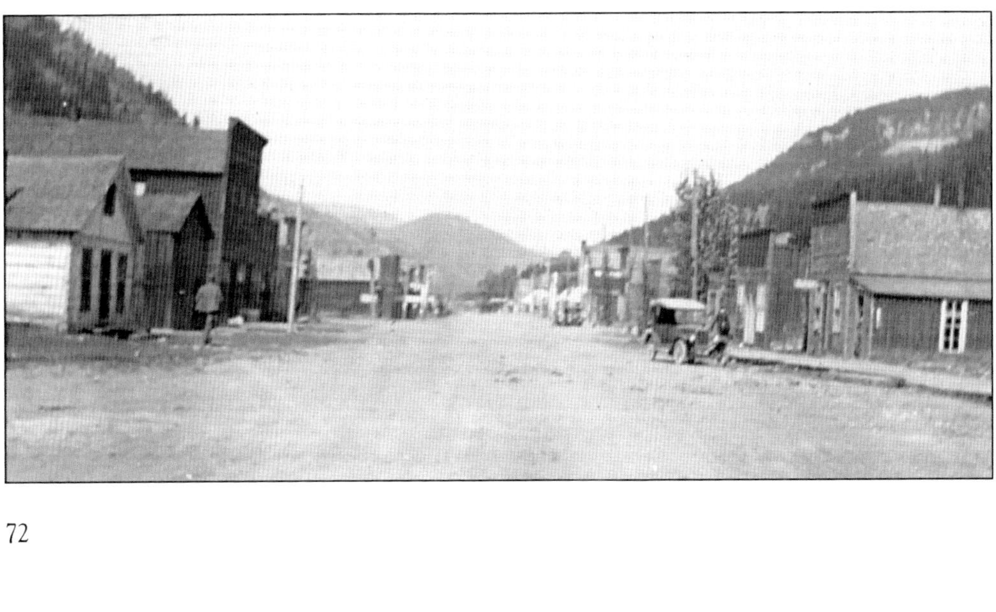

Four

Trails, Rails, and Roads

Sometimes you traveled in a wagon, at other times in a sleigh, and sometimes you were forced to walk. You had to push the wagon or the sleigh to help the fagged horses up the slippery hills, and by way of change you spent hours digging the almost smothered horses out of the soft snow or mending broken harness with rope, twine, or wire. It was the last straw on the camel's back to have to pay seven dollars for the privilege of riding on the stage.

— The Reverend J.J. Gibbons, *In the San Juan*

The remote location of the mining district was a factor that affected early development. Distance from the processing facilities and the lack of good transportation were some of the primary concerns. The transportation difficulties were somewhat alleviated with the construction of the Belt Mountain branch of the Great Northern Railroad, which connected Neihart with Great Falls in 1891. The smelter at Great Falls had been completed in 1888, providing a more accessible location for the processing of the district's ore.

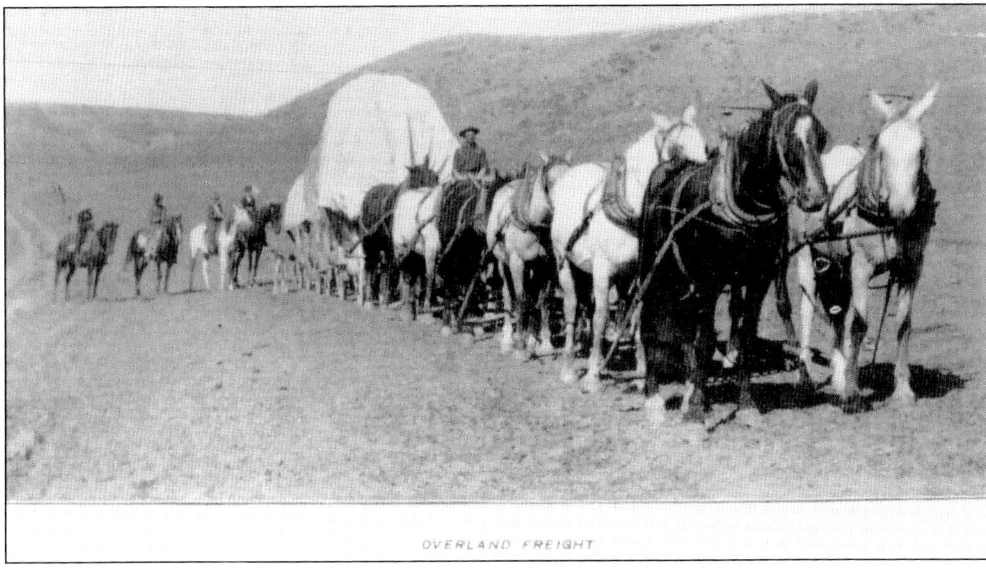

TRANSPORTATION PLAYS A KEY ROLE. Transportation served to reduce operating expenses and bring in machinery, equipment, and supplies. The first supplies to camp came by horseback over the trail from Barker in the spring of 1882. In August 1883, the first freight wagon train made camp. J.O. Hussey, of White Sulphur Springs, used a bull team to deliver supplies for M.L. Lohmire of the Massachusetts Mine. During the summer of 1883, a wagon road came from Belt Park down Harley Creek to Belt Creek, and camp was finally opened. (MOR.)

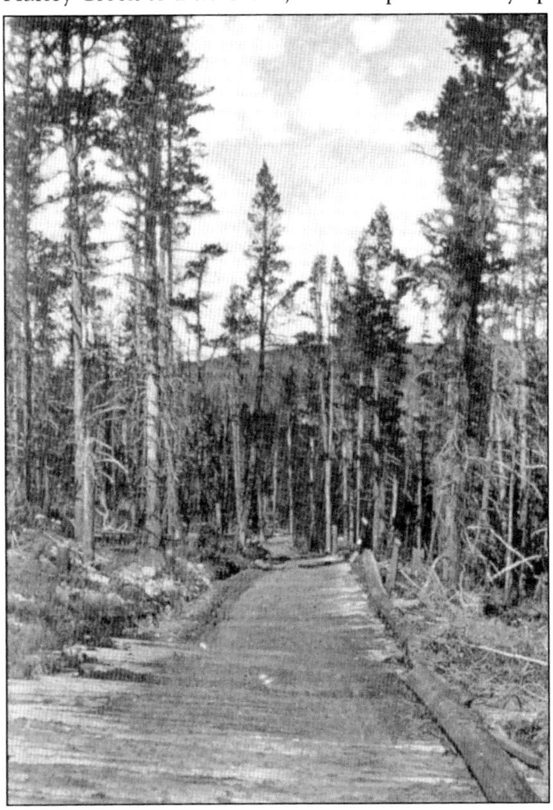

SAWMILL ROAD. The geology of the Little Belt Mountains is responsible for the breathtaking scenery, varied formations, and the mineral deposits that attracted miners and other pioneers to the area. Miners and merchants created this tree-lined wagon road along Sawmill Creek heading toward Wolsey and White Sulphur Springs. (Tony Juntunen.)

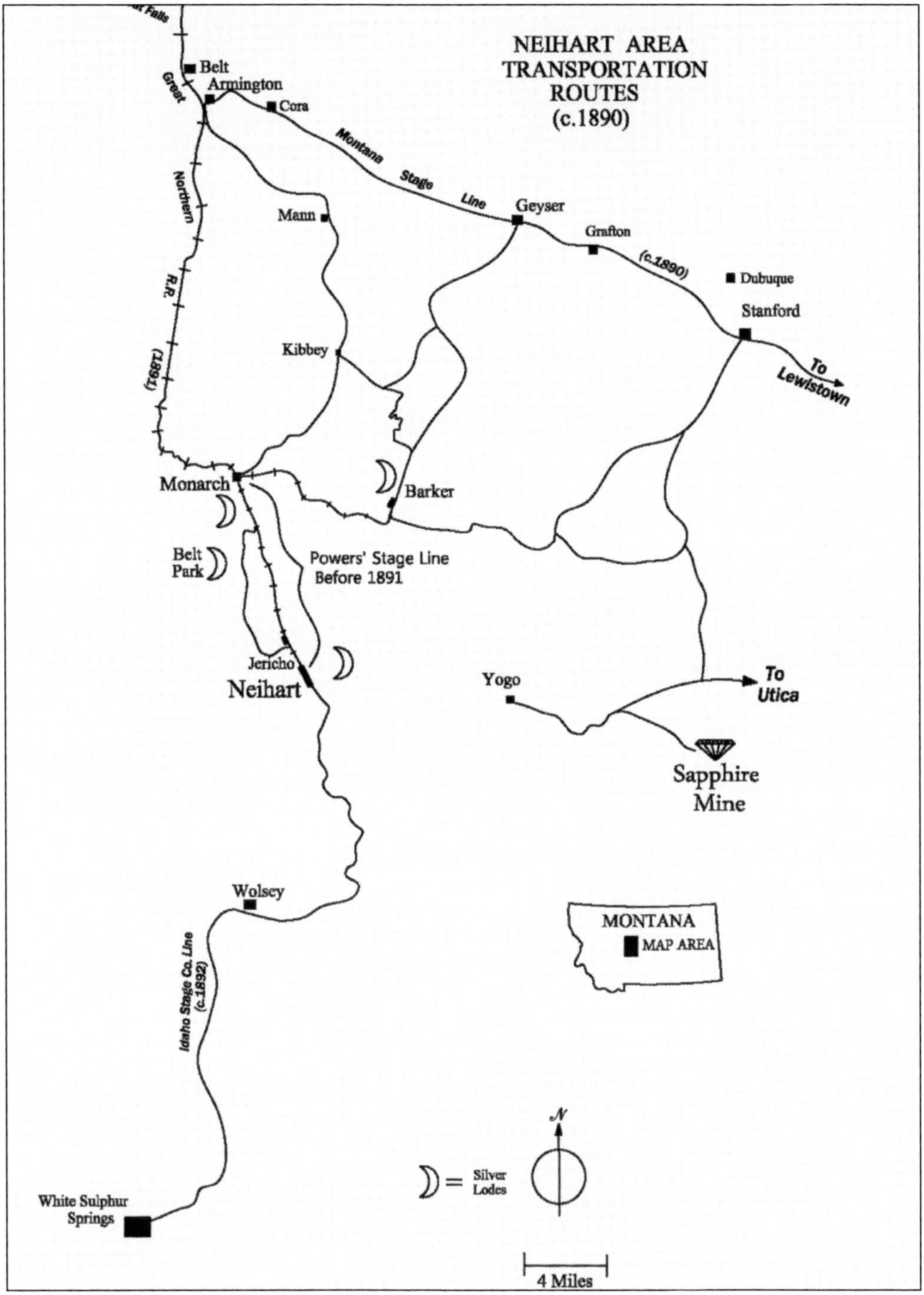

Little Belt Mountains Transportation. Based on historical documents, newspaper accounts, and other books and articles on the Little Belt Mountains area, this map depicts travel routes during the initial days of Neihart's mining activities. A few wagon roads existed prior to the settling of Neihart—some were created because of the foresight of J.L. Neihart, and some were adopted from Native American routes. (Author's collection.)

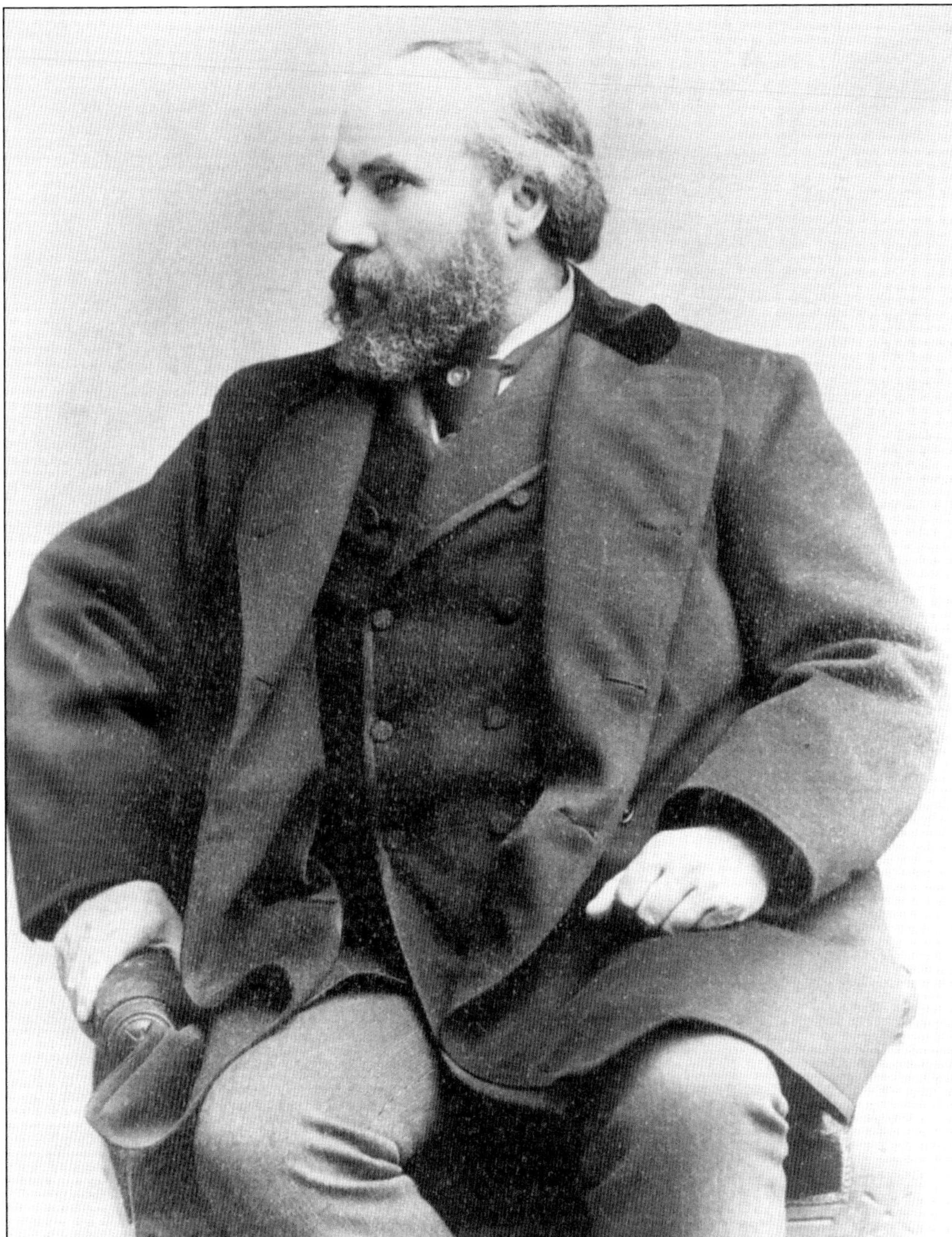

James Jerome Hill, c. 1879. Montana businessman and "empire builder" James J. Hill wanted to build an empire that would control transportation and communication from the Mississippi River to the Pacific Ocean. His goal was to undercut the Northern Pacific Railroad by offering lower prices. Hill's transcontinental railroad did not receive land grants, as the Northern Pacific did. Instead, he attracted big private investors and, with their support, Hill built the Great Northern Railway, stretching from Minnesota across North Dakota, Montana, and into Washington. The route became known as the High-Line because it was the northernmost transcontinental route in the United States; it also refueled Neihart's life in the early 1890s. (MnHS.)

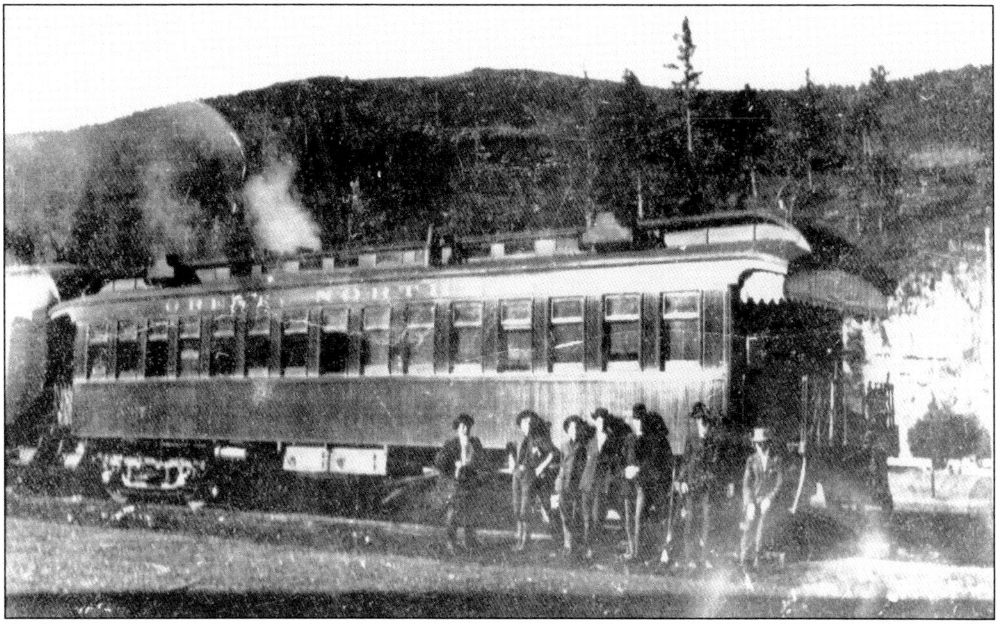

Great Northern Railroad Survey Car. According to the brief notation discovered with the photograph, the rail survey car was on a siding in or near Neihart. The subjects leaning against the car appear to be younger than typical area miners. Based on the hiring practices of the Great Northern Railway in Saint Paul, Minnesota, James Hill demanded quality work by quality people. For engineering-type positions, he was a stickler for education. So in all probability, these surveyors were fresh-faced college men. (THM.)

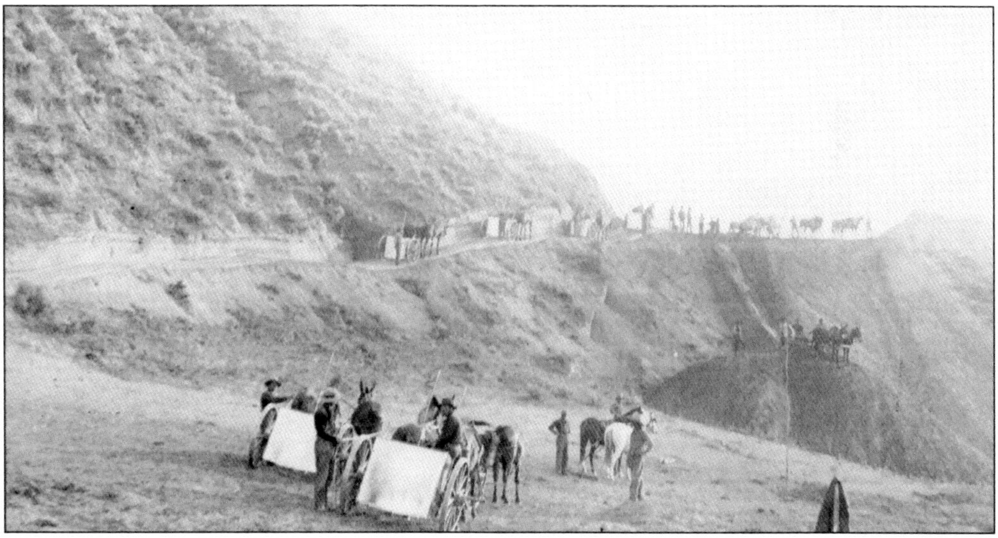

Crews Grading and Hauling Iron Between Belt and Neihart, c. 1890. The remoteness of mining districts impacted early development. The distance of the Neihart mining district from the processing facilities and the lack of good transportation were some of the primary concerns. The transportation difficulties were somewhat alleviated with the construction of the Belt Mountain branch of the Great Northern Railroad connecting Neihart with Great Falls in 1891. The smelter at Great Falls, completed in 1888, provided a more convenient location for the processing of the district's ore. (James J. Hill Papers, MnHS.)

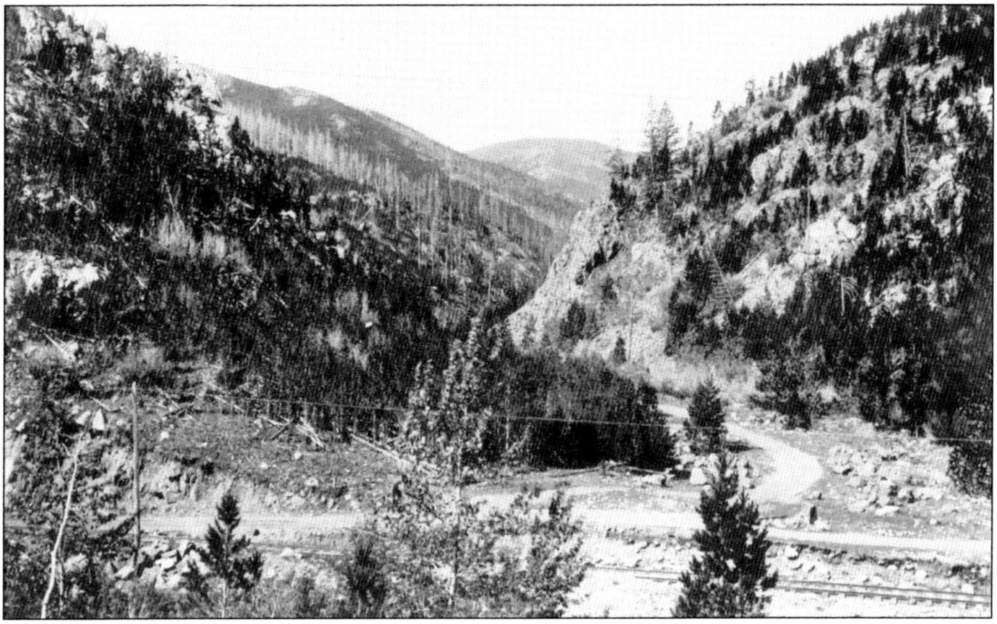

HARLEY CREEK BRIDGE LOOKING SOUTH UP HARLEY CREEK, C. 1890. Also taken from the vantage point of the Hudson Mining Company smelter, this photograph shows Harley Creek winding south from the southern banks of Belt Creek. Three mining operations on Harley Creek were the Imperial Group, the Royal Group, and the Granite Mining Company. There used to be a graveyard at the head of the gulch, so it was sometimes called Graveyard Gulch. All remains were reportedly reinterred elsewhere before the end of the century. (THM.)

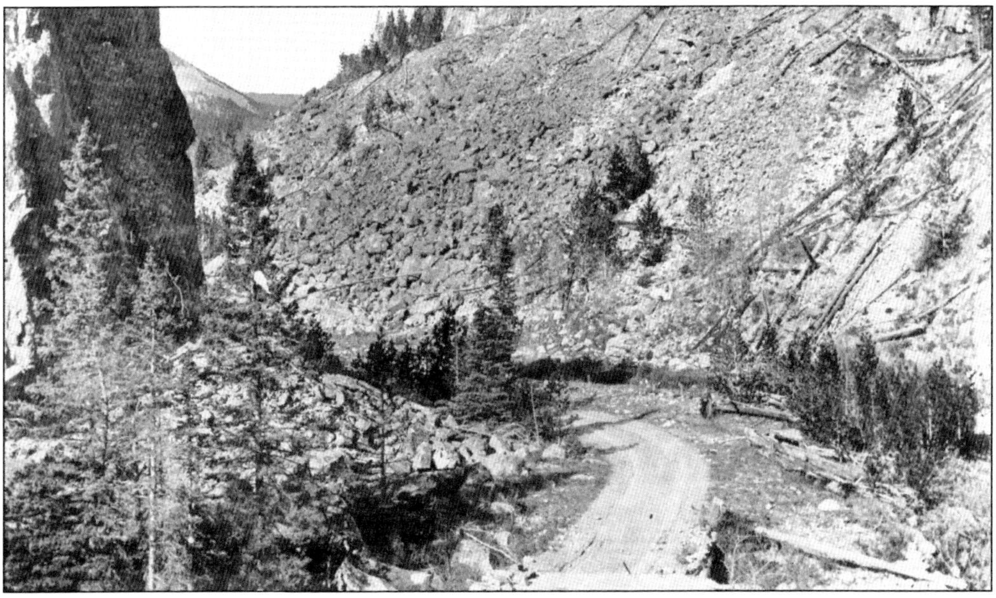

HARLEY CREEK BRIDGE SHOWS EARLY NEIHART ROAD AND GULCH RUGGEDNESS. The log bridge in the foreground crosses Harley Creek (named for Richard Harley, one of the three discoverers of the Neihart mining camp) as it flows into Belt Creek to the right. The Harley Creek District, three miles down from Neihart, was known for its large veins of copper ore, described in the *Neihart Herald* as being "of monstrous size, some standing 80 feet from wall to wall." (THM.)

MONTANA CENTRAL RAILROAD'S ORE-LOADING DOCK, 1893. Some mines in Neihart were fortunate enough to not require the use of this ore-loading dock. However, the ore from a mine not located next to the tracks had to be hauled to the rail yard for transportation to Great Falls and processing. Visible to the left in this photograph is a horse barn and stable near the area of the third school building in Neihart. Built in 1920, it is listed in the National Register of Historic Places and now serves as the senior citizens' center. Behind the railcar appears a structure seen in nearly all early photographs of Neihart—the first home of town attorney and mine owner Richard Bennett, which stills stands today. (Dick Bennett.)

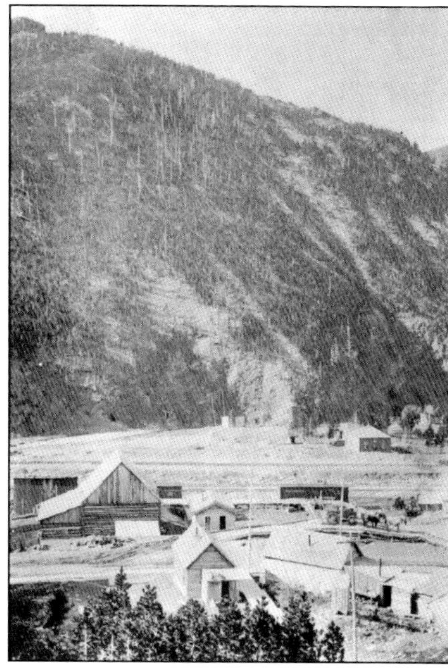

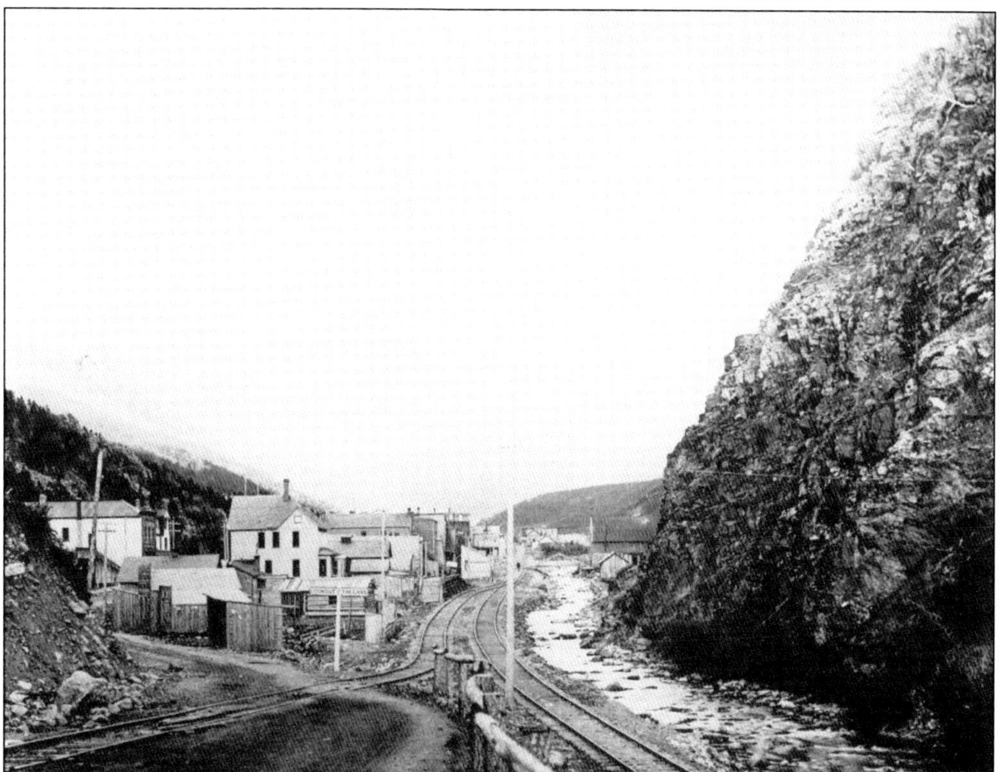

ROUNDING THE CURVE INTO NEIHART. This is the view into Neihart from Monarch. The guardrail marks the right-of-way of the Queen of the Hills and a public road. Where the road crosses the tracks stands a sign warning, "lookout for the cars." This road heads up towards Rock Creek and eventually winds its way down onto Neihart's Main Street. (THM.)

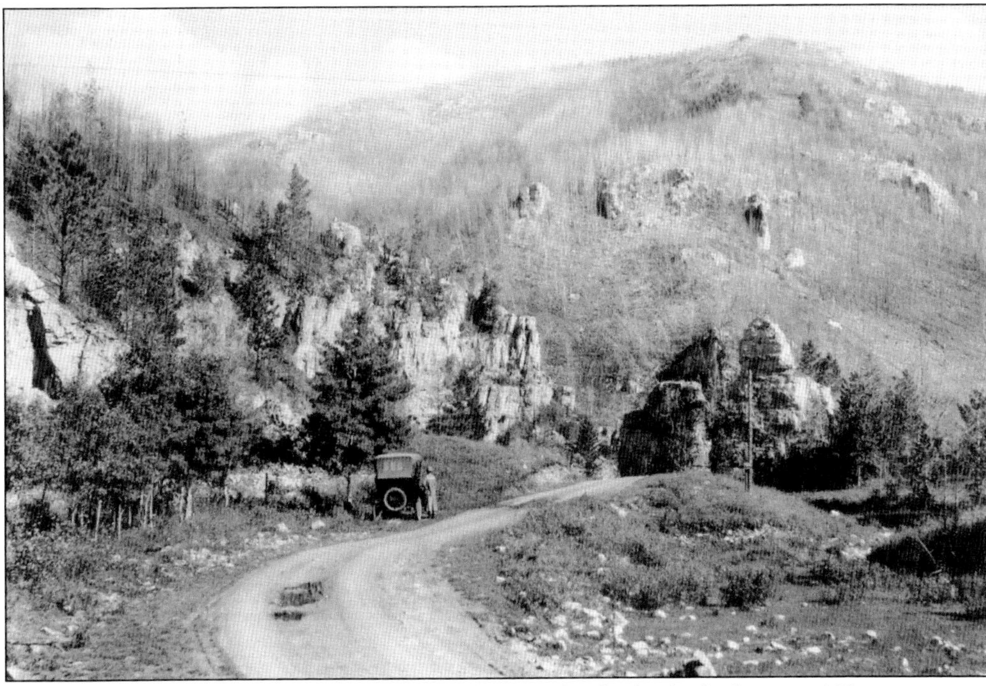

Trails, Automobiles, and Scenery. These photographs display both the incredible area scenery and another engineering challenge: roads. It was not long after trains reached Monarch, Barker, and Neihart that the automobile arrived in the mountains. With it came the need for roads in better shape than the old wagon trails. The exact location of these photographs in the Little Belt Mountains is unidentified, yet both show incredible scenery and the effects of erosion. These images also offer a reason why US Highway 89, between White Sulphur Springs, Neihart, Monarch, Riceville, and Armington Junction, was named a National Scenic Byway by the Federal Highway Administration, "connecting Central Montana's silvered past with its bright recreational future." (Both, THM.)

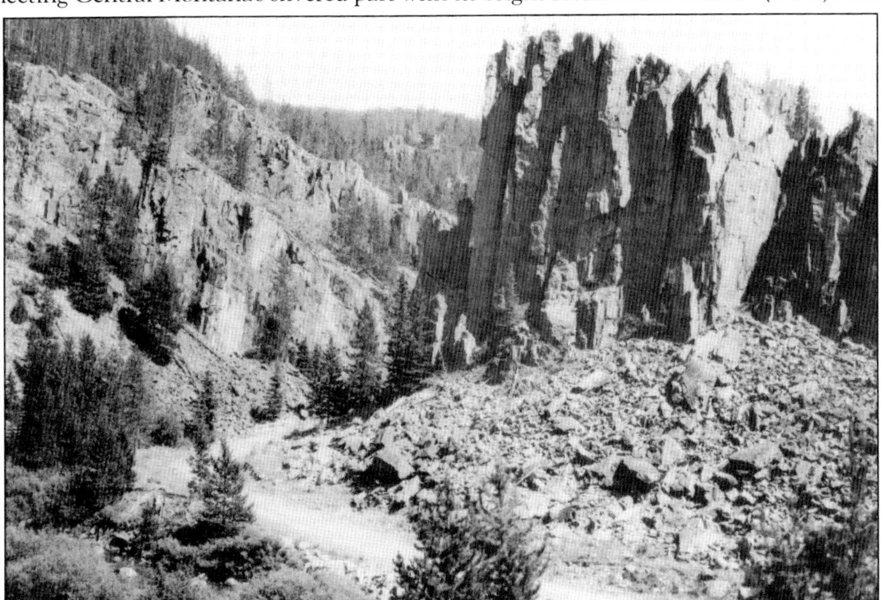

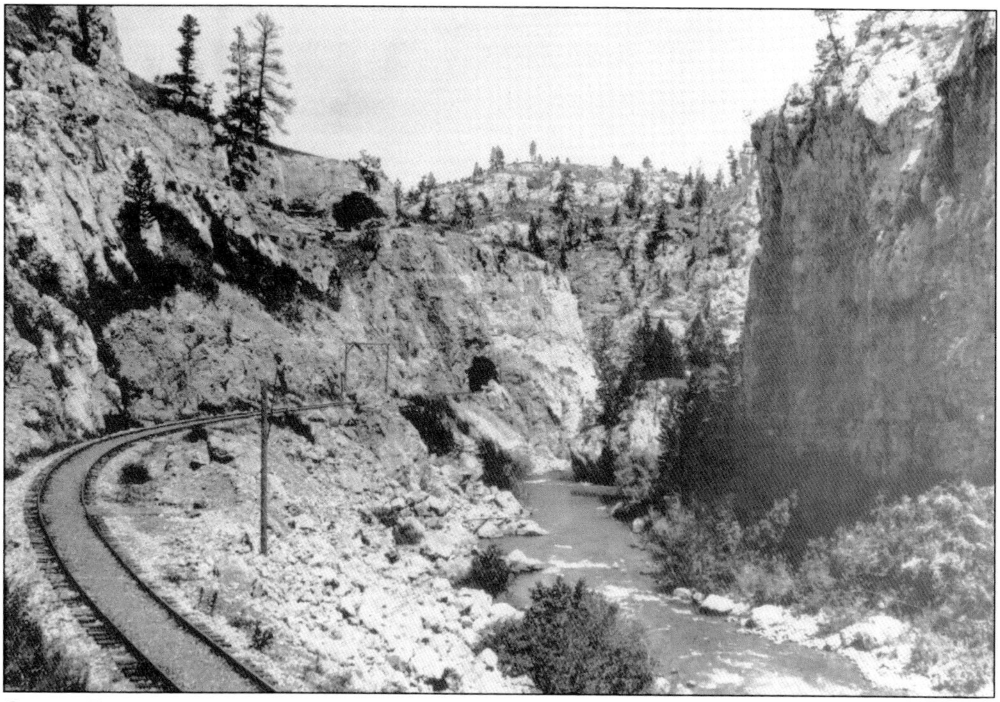

Creek, Rocks, and Tracks. These photographs portray the scenery and the engineering feat of the Great Northern Railroad of Saint Paul, Minnesota. Using narrow-gauge tracks and small-sized railcars to accommodate the tight curves and tunnels, crews laid tracks, and laborers created tunnels and bridges. These men presented a cross section of nationalities similar to those of the miners in Neihart and other camps. They worked together, apparently harmoniously, to construct the bridges, trestles, and track from Great Falls to Neihart. The dark opening visible in the tough limestone wall marks the location of one of many long tunnels that allowed the 50 miles of track from Belt to reach Neihart and the ore. The photograph below provides a closer view of the tunnel and its length. (Both, THM.)

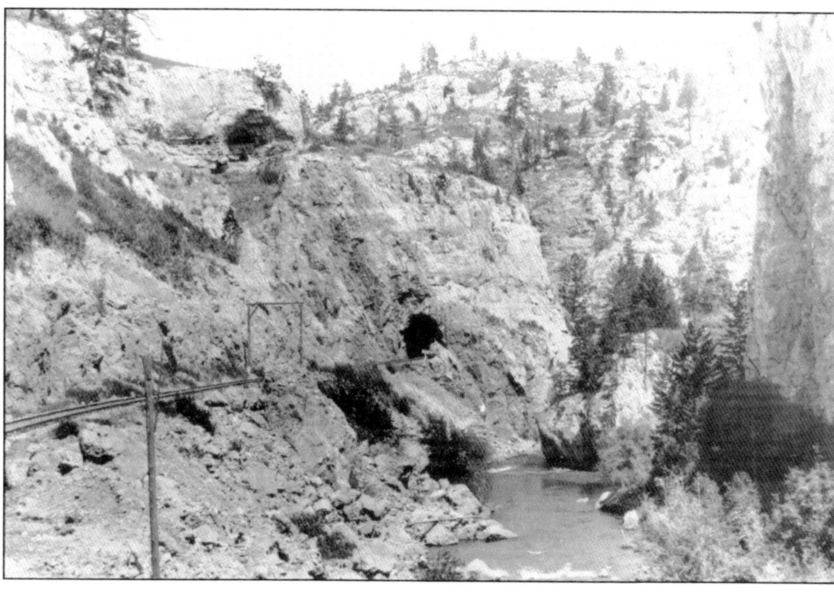

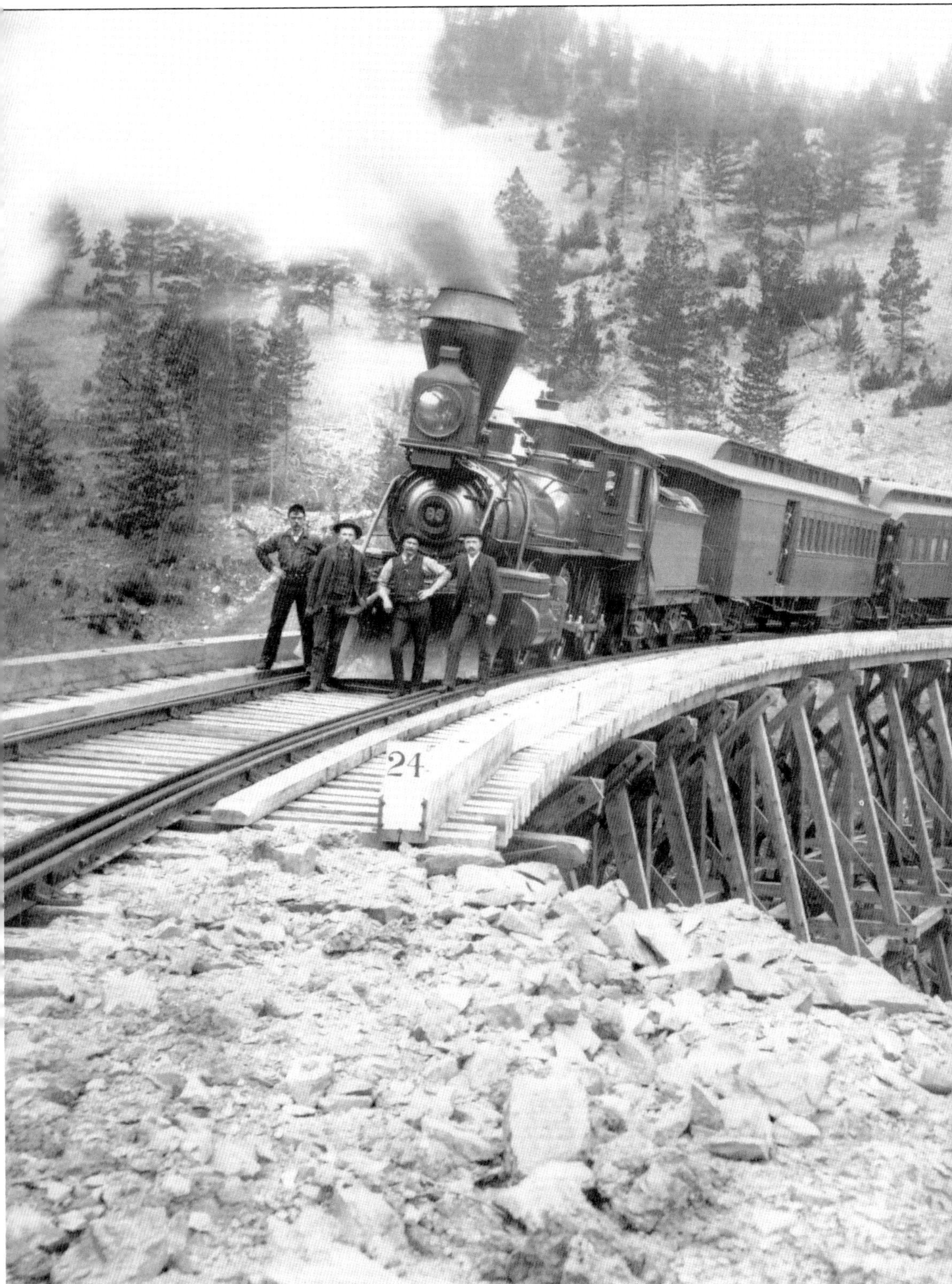
24

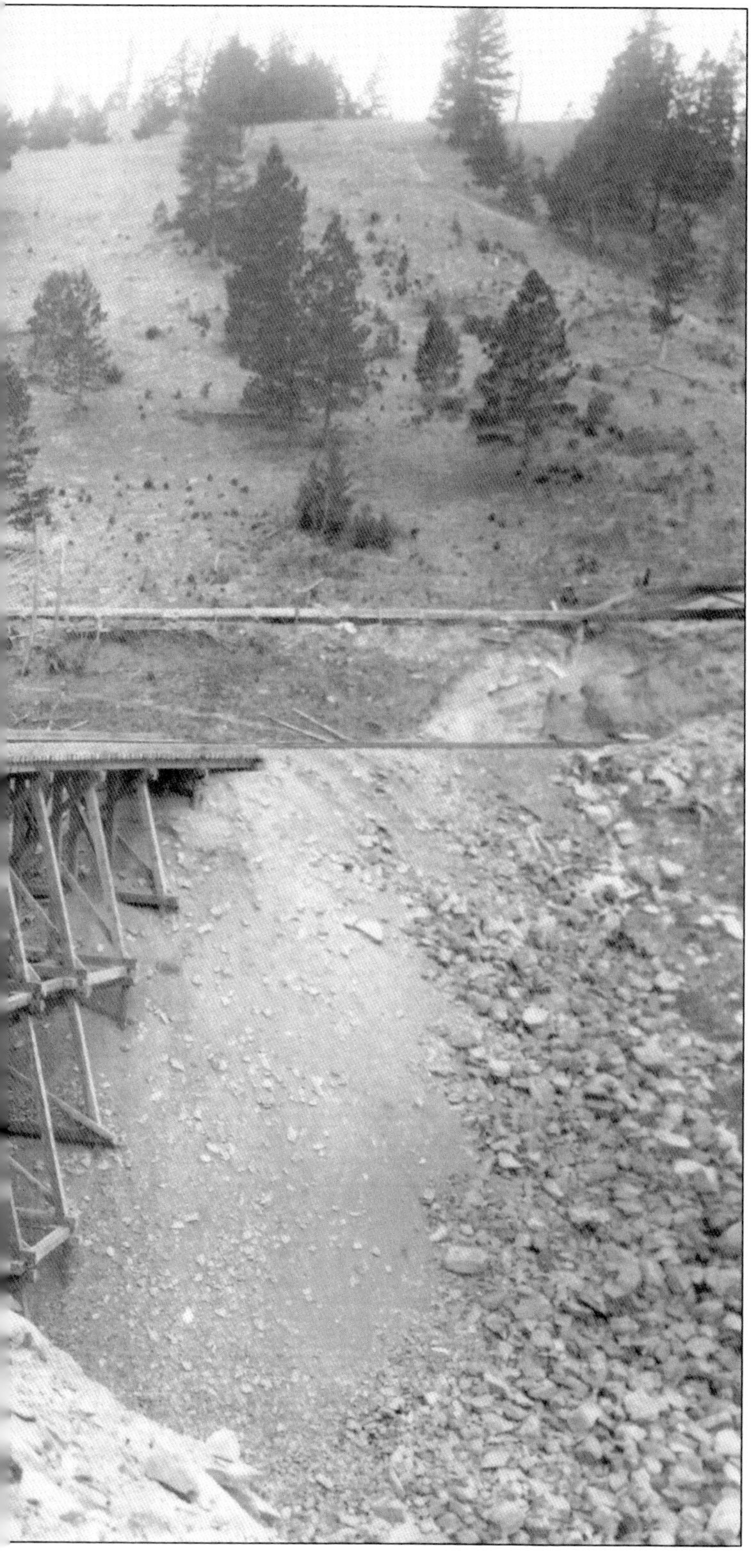

NEIHART FLIER HIGH ABOVE A GULLY. This undated photograph is estimated to have been taken sometime between 1891 and 1895. The Neihart Flier consisted of the engine, coal and wood car; baggage, mail, and passenger car; and a full passenger car. The individuals standing in front of the snowplow are unidentified, as is the gentleman wearing a bowler standing in the baggage and mail door. At the front of the last car, two more unidentified persons appear, with one standing on the bridge and the other on the platform. Just visible above the rear of the train appears to be a wooden water flume. The bridge was numbered 24, as seen on the printed sign on the lower right bridge beam. (THM.)

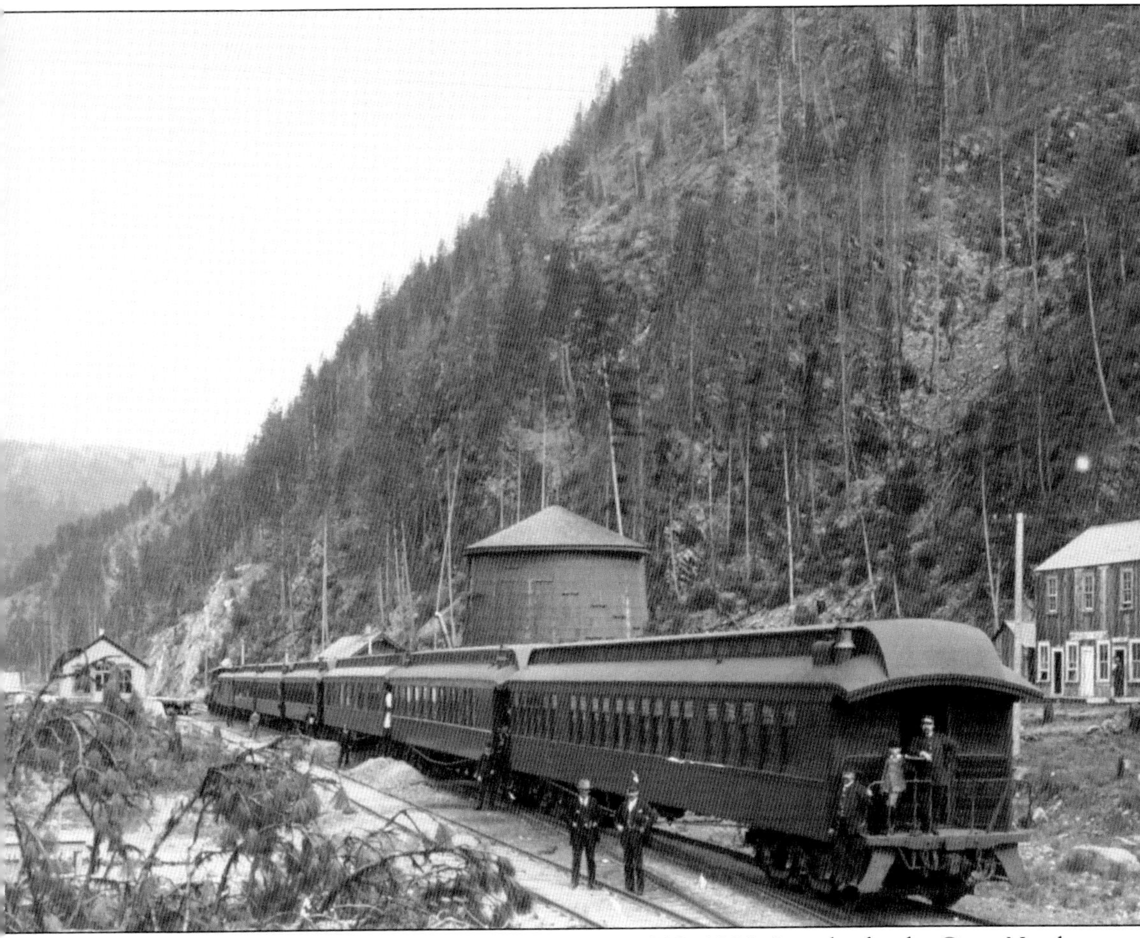

IMPRESSIVE ARRAY OF COACHES IN MONTANA, C. 1893–1894. In order for the Great Northern Railroad to capitalize on the transportation service it offered to miners, farmers, ranchers, merchants, and others with freight-hauling needs, it needed to provide a full-service railroad. The Great Northern was not a part of the land grant provided to earlier railroad expansion plans. The Pacific Railroad Act of 1862 provided alternating 15-mile parcels of land on each side of the tracks, allowing the railroads to arrange financing through land deals. James J. Hill had to finance his railroads through commerce. (James J. Hill Papers, MnHS.)

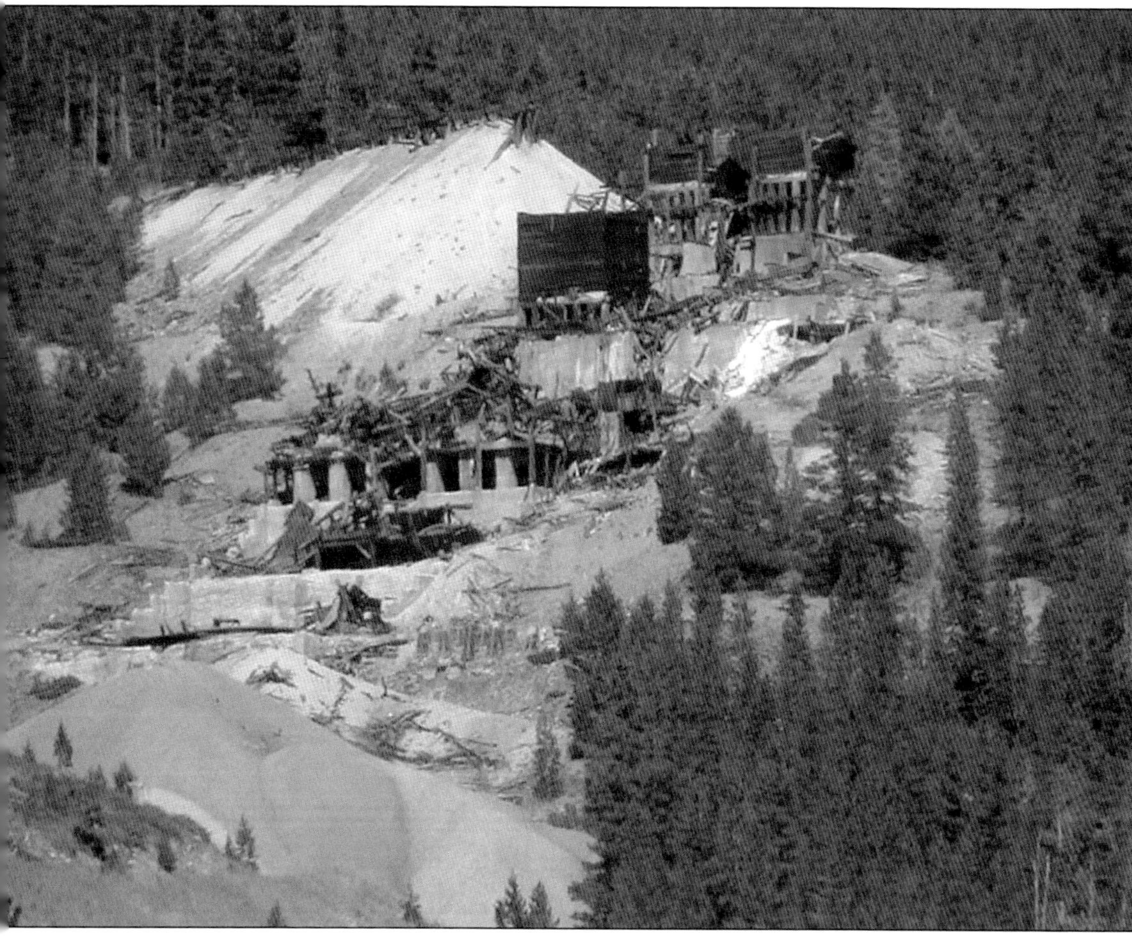

Remains of the Silver Dyke Mine. Silver price fluctuations created booms and busts for the Neihart mining district's history. Following World War I, prices again rose, as did the fortunes of the Silver Dyke Mine. The more it produced, the smaller the space for its tailings, with the company eventually dumping directly into Carpenter Creek, adding to the pollution of area streams and conflicts with other mountain residents. Reports of spoiled wells and cattle deaths added to the growing problem. On July 10, 1925, a suspected result of an earthquake, the Silver Dyke's tailing pond either burst or collapsed, sending tailings and sediment roaring down Carpenter Creek into Belt Creek and coursing its way to the Missouri River. Two children, asleep in their home in the mining camp, were swept to their deaths that night. This photograph provides a visual idea of the size of the Silver Dyke's operations. (Author's collection.)

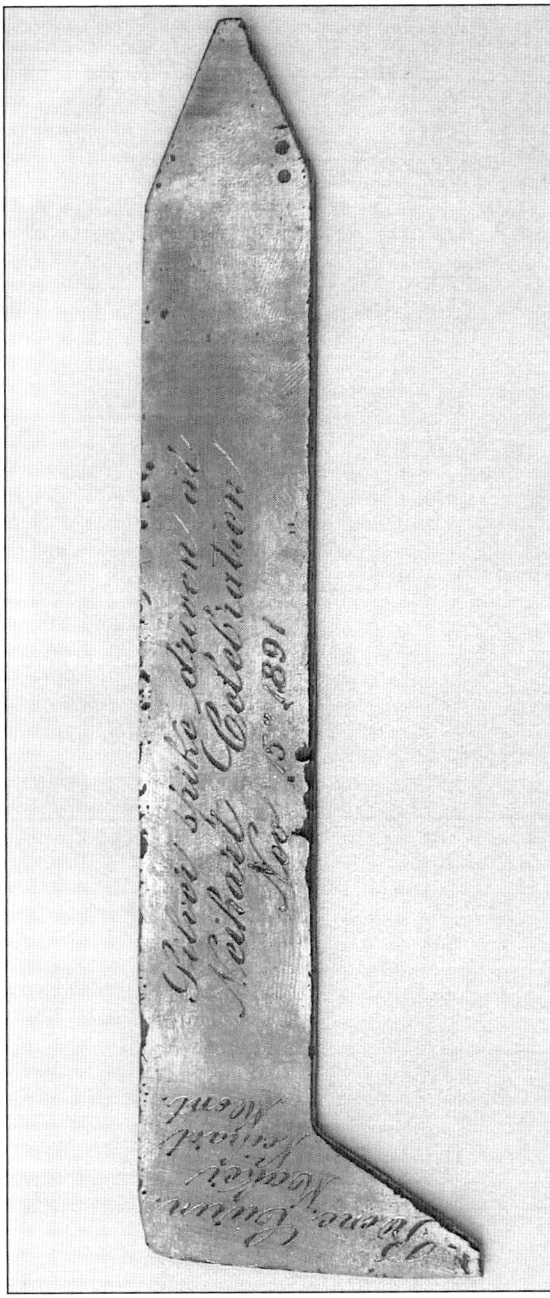

MONTANA CENTRAL RAILROAD ARRIVES IN NEIHART, NOVEMBER 15, 1891. Finally, on November 15, 1891, it happened; the long-awaited, anguished-over, and sorely needed had arrived—transportation for the ore, the goods, and the people. The celebration was something to write about, and write they did. The *Great Falls Tribune*, *Neihart Miner*, *Neihart Herald*, *Saint Paul Pioneer Press*, *Rocky Mountain Husband*, and *River Press* all reported, "On the 15th day of November, 1891, the silver spike was driven by Miss Mabel Brennan, daughter of Richard Brennan, one of the first family men in Neihart." This is a photograph of the engraved spike, made from silver mined from the Queen of the Hill Mine. (THM.)

Five

Geology and Scenic Wonders

People of today do not realize how hardy men had to be to live in a steamy, smelly bunkhouse with 200 or more grimy men in damp clothes coming and going continuously with hob-nailed boots and with no place to go other than down that wet shaft, to the bunkhouse and boardinghouse or outdoors into 10 or 12 feet of snow slides and work for 10 hours with heavy hand tools.

—J.T. Pearson

Prior to the arrival of white settlers and miners, the area north of today's Monarch was a passageway to the Great Plains for native buffalo hunters. Tepee rings can still be found in remote corners of Meagher and Cascade Counties, where nomadic tribes camped and chased the great herds. In the 1860s, the sparkle of gold brought hordes of prospectors to the territory, forever changing the landscape of Montana. Of course, the discovery of silver several decades later allowed Neihart to burn brightly as one of the area's boomtowns, if for only a short while.

The Little Belt Mountains were a paradise running through the heart of Montana, with richly diverse landscape, wildlife, history, culture, and commerce. This region possesses qualities unmatched by any other patch of the West. Neihart Canyon combines rural charm and Western sensibilities with worldwide appeal.

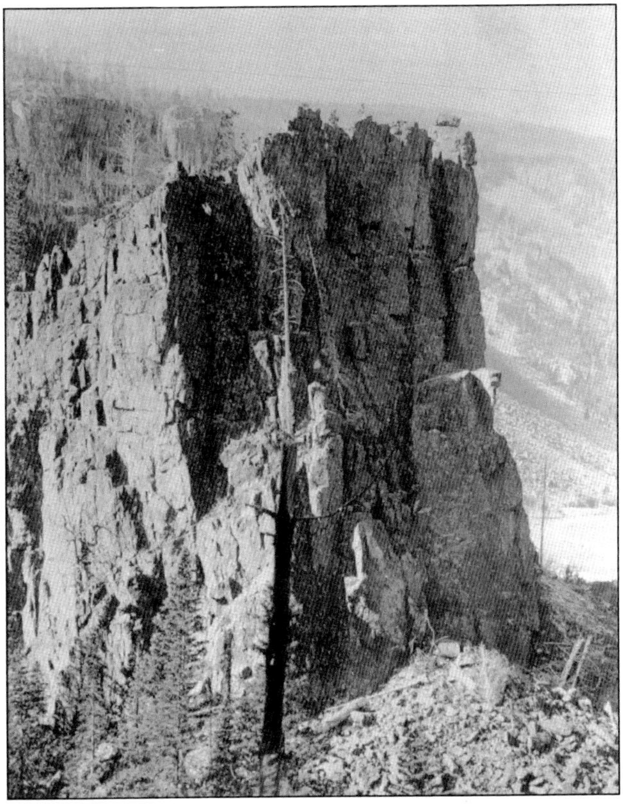

Beauty and . . . Two stunning photographs, taken around 1893, reflect the scenery that can still be enjoyed by people traveling through Montana's Little Belt Mountains. The effects of erosion on the Belt Creek Canyon (or Neihart Canyon) draw frequent comments from visitors surprised to find such beautiful scenery on their trips to explore old mining towns. (THM.)

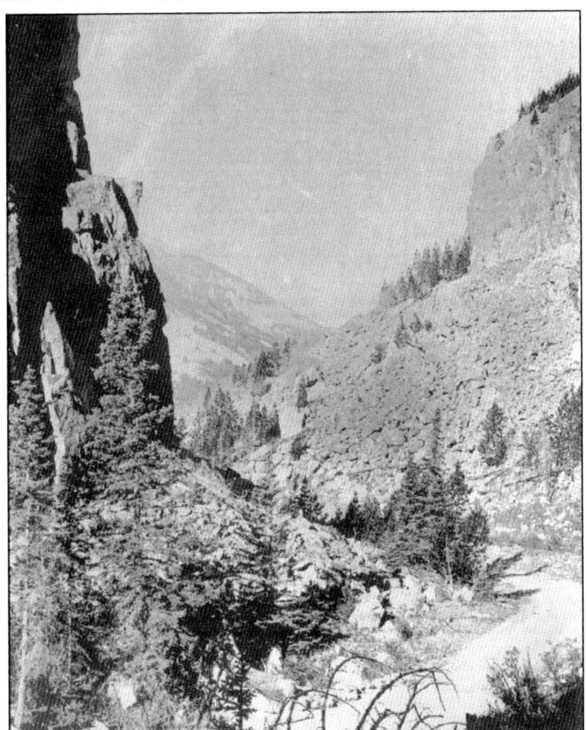

. . . Ruggedness. Some locals suggest that these two photographs are of an area known as the Devil's Chair, which can be found by turning left off Highway 89 south of Neihart, just past Memorial Falls, onto Jefferson Creek Road. This impressive display will be on the left. (THM.)

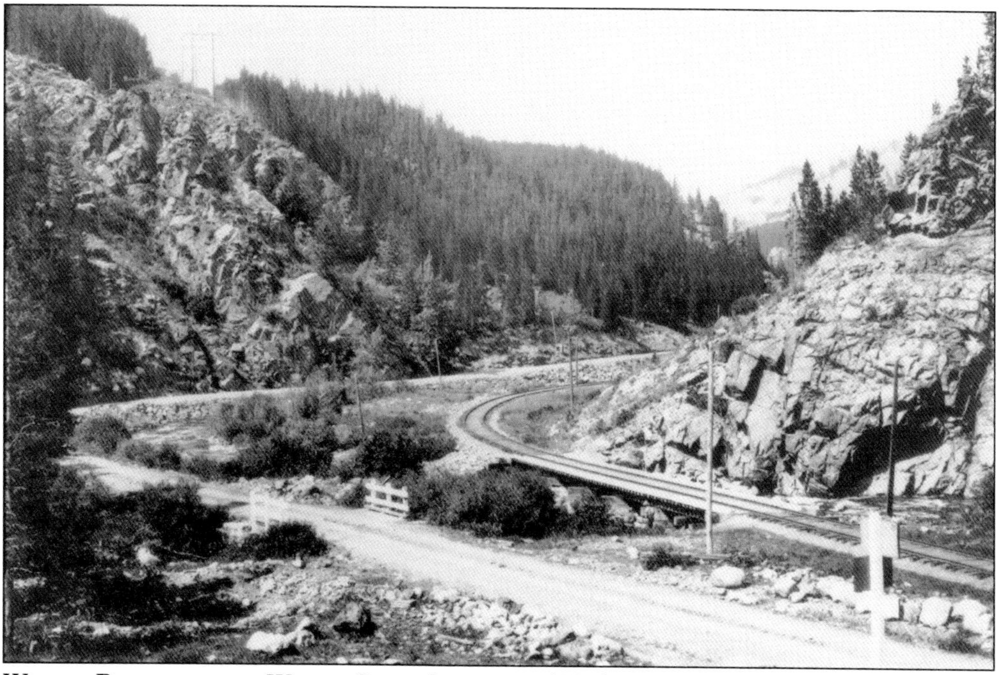

Wagon, Railroad, and Water. Somewhere around Neihart or Monarch, there appears a hard turn to the right where all three modes of transportation exist together alongside the surrounding nature. In the upper left of the photograph, power lines are visible (either for telegraph or telephone) alongside the rail tracks, as well as a dirt road, traffic sign (lower right corner), and bridge over Belt Creek. The important point of this photograph is little Belt Creek, which, over millions of years, has cut a path through these mountains, wearing away the shale shown alongside the tracks and under the power lines, eventually meeting with the Missouri River below Belt. (THM.)

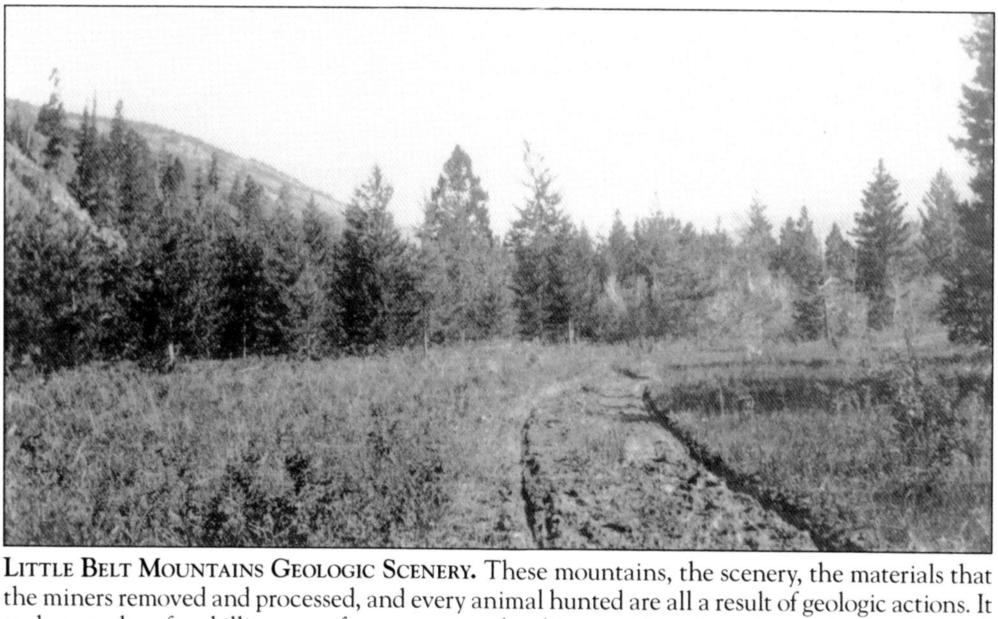

Little Belt Mountains Geologic Scenery. These mountains, the scenery, the materials that the miners removed and processed, and every animal hunted are all a result of geologic actions. It took more than four billion years for nature to sculpt this area. Nature's work was slow, methodical, and bruising. (THM.)

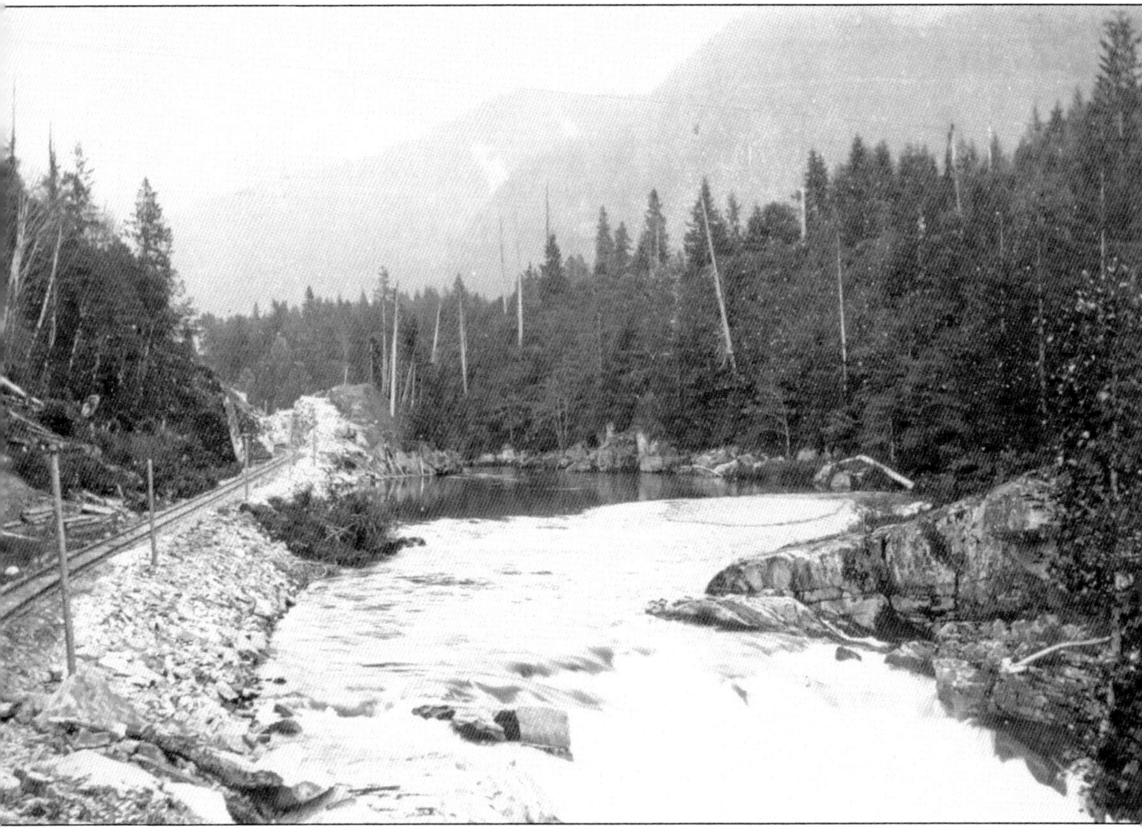

Beautiful Scenery Enjoyed by Those Riding the Train to Neihart. One of the highlights for any Monarch, Barker-Hughesville, and Neihart passenger was the beauty of the view from the train as it passed through the canyon of Belt Creek. The Little Belts (as the locals call them) were formed by a dome of lava running about 80 miles north to south and around 30 miles west to east. This landscape is a stunning example of what the pioneer men, women, and families saw, endured, and overcame. (James J. Hill Papers, MnHS.)

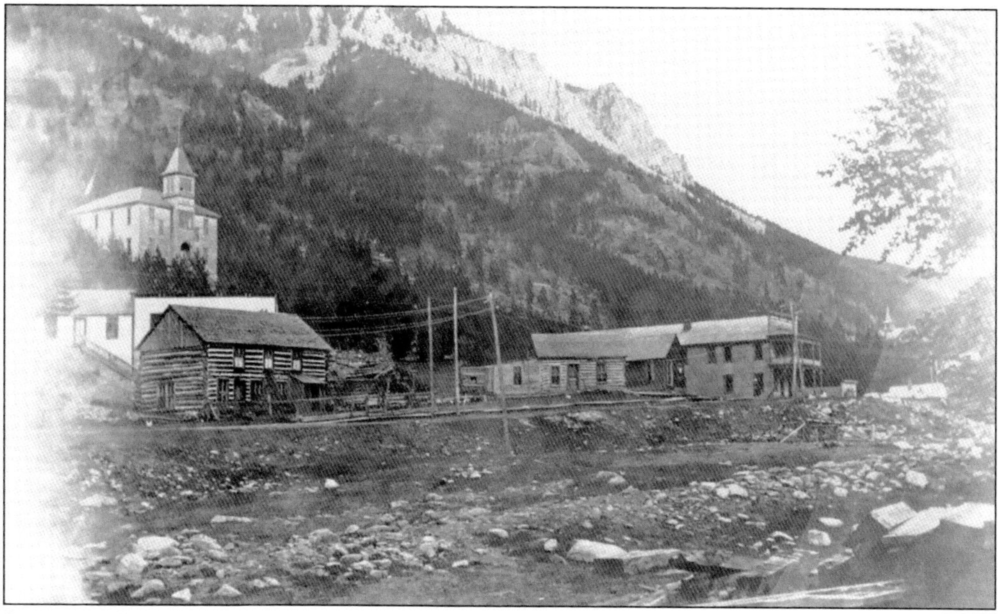

Striking Outcrop on Neihart Baldy. Apparently, nobody staked a claim on that distinctly shaped outcropping that highlights the skyline above south Neihart. Bull of the Woods Mine was up at that elevation but on the other side of Baldy. It is thought that this piece of natural beauty remained untapped. Down below, however, much more is known. After this photograph was taken in October 1891, the ore-loading dock, horse barn and stable, and railroad tracks were eventually added. The school built in 1891 stands sentinel over the log two-story home of Harry D. Burghardt, with the corner of John Workman's home visible. To the right are the Neihart Hotel (with second floor balcony), and the Pete Rush–Henry Sutton house. The crop of rocks in the foreground shows the challenge of digging in Neihart. (THM.)

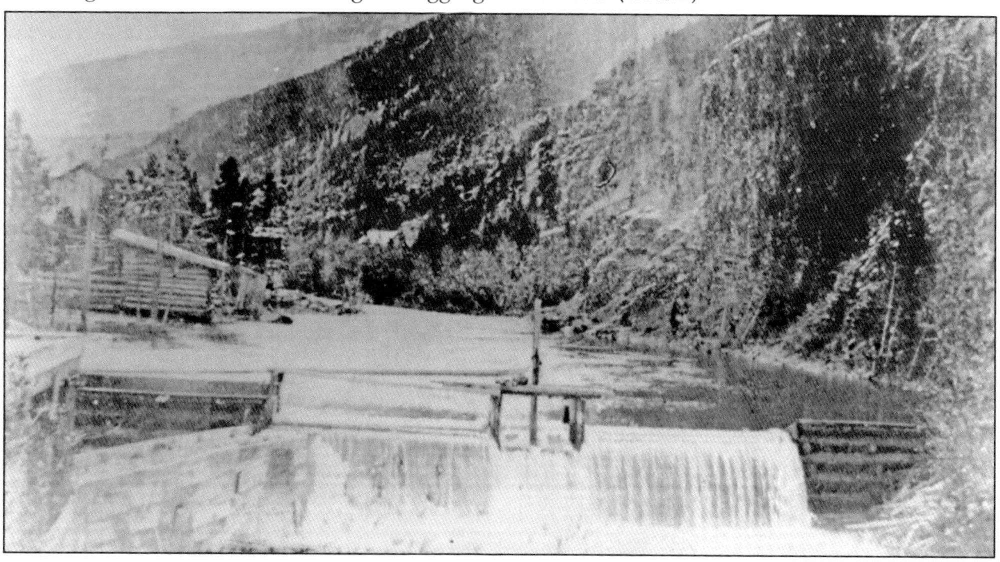

Water, Water, Water. The many dams created on the Belt Creek, like the one pictured above, were first constructed to provide power to turn mill wheels, generators, and small grinders. One encountered these power dams up and down the entire one-and-a-quarter-mile length of the Neihart mining camp. (THM.)

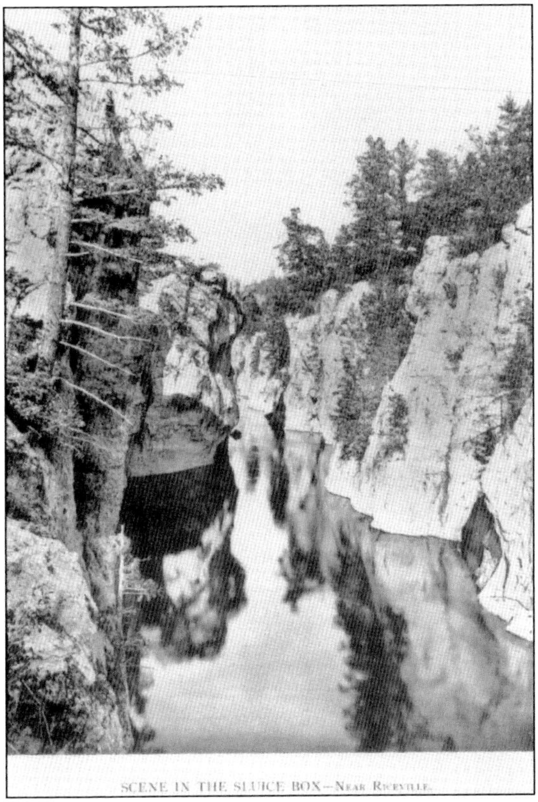

SLUICE BOXES STATE PARK. This little-known state park has been called a "Grand Little Canyon" for good reasons. Jeff Erickson described it, in a 2006 article for *Montana Outdoors*, as "a pretty kid sister to the Smith River canyon." The name refers to the canyon walls, which resemble the vertically walled sluice boxes used by early-day placer miners, such as those from Virginia City or Last Chance Gulch. Remarkably, the Montana Central Railroad (a branch of the Great Northern Railroad) did some serious engineering through this canyon area to get to Monarch and then up to Neihart in the early 1890s. (Both, THM.)

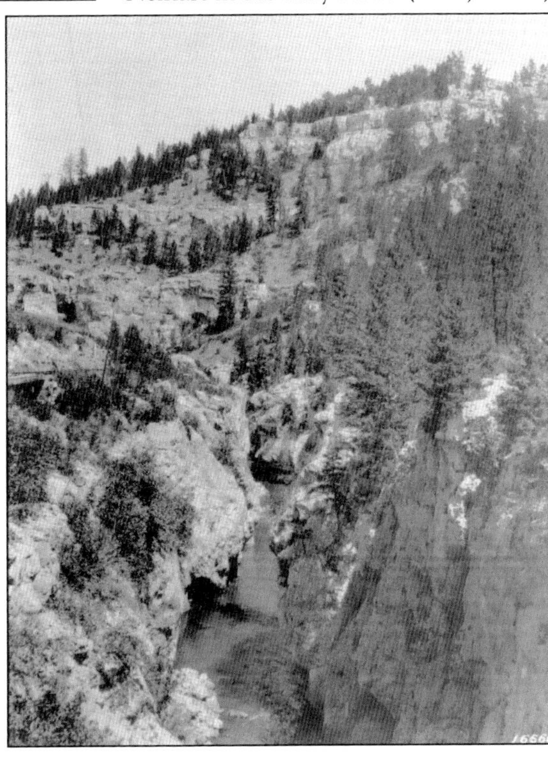

Daniel Dutro Photographs Scenery in Little Belt Mountains. Dutro found his way to Montana Territory after the Civil War and after spending a couple of years working with the Vigilante Committee in Missouri, tracking down Confederate troublemakers. In 1867, he traveled on a river steamboat to Montana, arriving in Fort Benton and then heading to Helena. In order to be outside daily, he headed for Neihart by way of Confederate Gulch and White Sulphur Springs and established the Benton Group of mines. Eventually, he moved to Fort Benton, where he purchased a photography studio and discovered his talents as a photographer. (MHS; photograph by Daniel Dutro.)

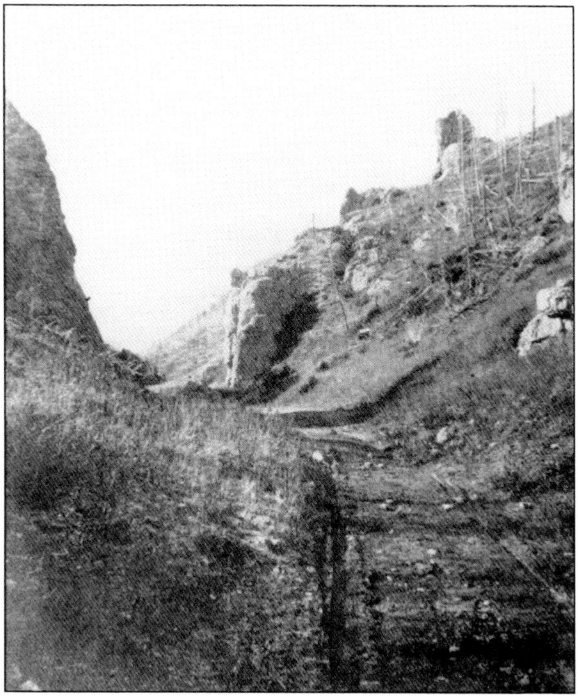

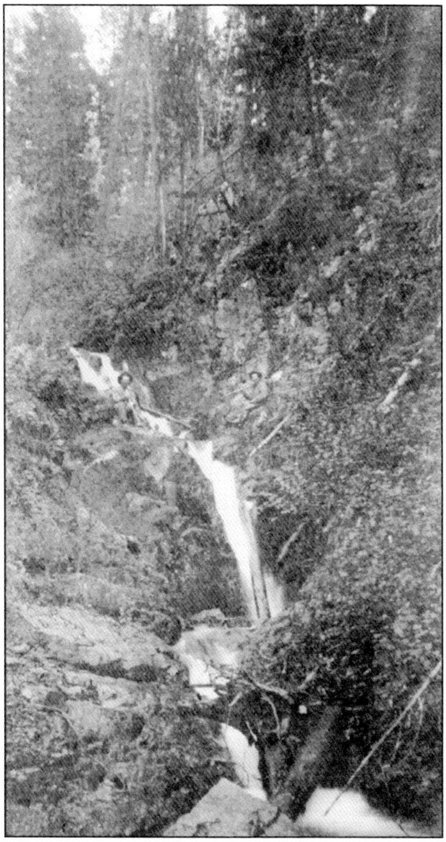

Little Cascade. This photograph, titled "Little Cascade in Belt Mountains," features a hefty waterfall on an unidentified creek somewhere in the Little Belt Mountains. The second is titled "Scene in Belt Park." It is a good view of the road between Monarch and Graveyard Gulch. Daniel Dutro's true calling was good fortune for Montana since he produced a wealth of unique photographs capturing those early days of Montana's history. (MHS; photograph by Daniel Dutro.)

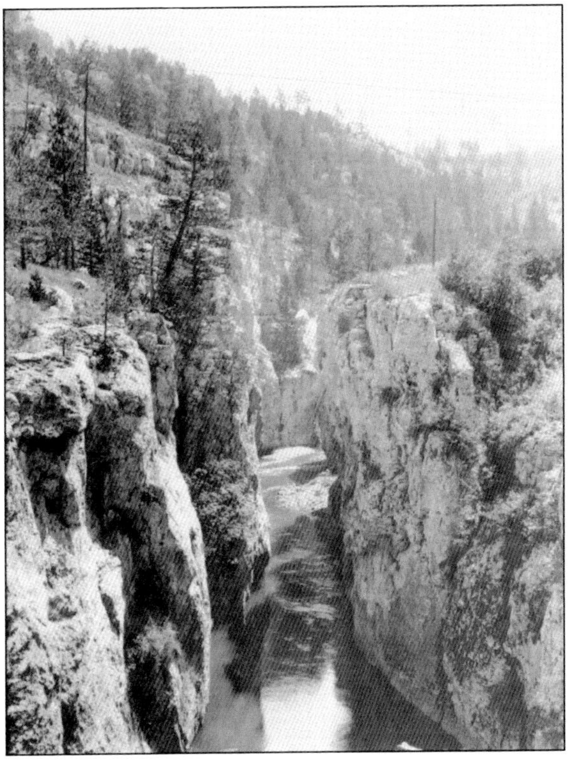

Belt Creek behind Monarch. This photograph captures how, over time, flowing water carves a passage through hundreds of feet of rock. Notice how the creek meanders—right, left, right, straight for a while, left—as it continues to wind its way downhill towards its eventual ending in the Missouri River. Whenever flowing water hits that limestone canyon wall, a little piece of the limestone flows away, with the stream cutting into and altering the course of the flow and the size of the wall. Over time, flowing water always wins. (THM.)

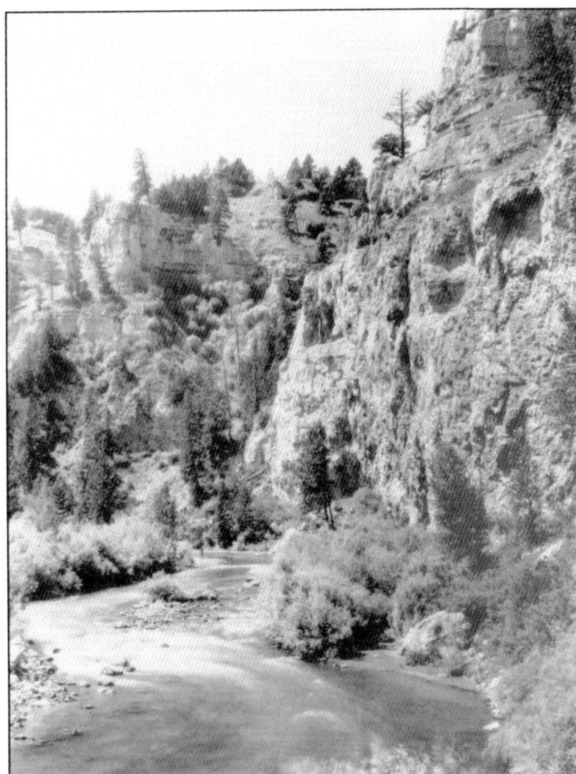

No Match for Flowing Water? What an awesome photograph of a part of the Little Belt Mountains that few travelers have ever seen. Neatly incised between the walls of Madison limestone, the Little Belt Creek carves a small passage through the beauty of the Sluice Boxes State Park. How logs were successfully floated down Belt Creek to a sawmill through this narrow slot continues to be baffling. (THM.)

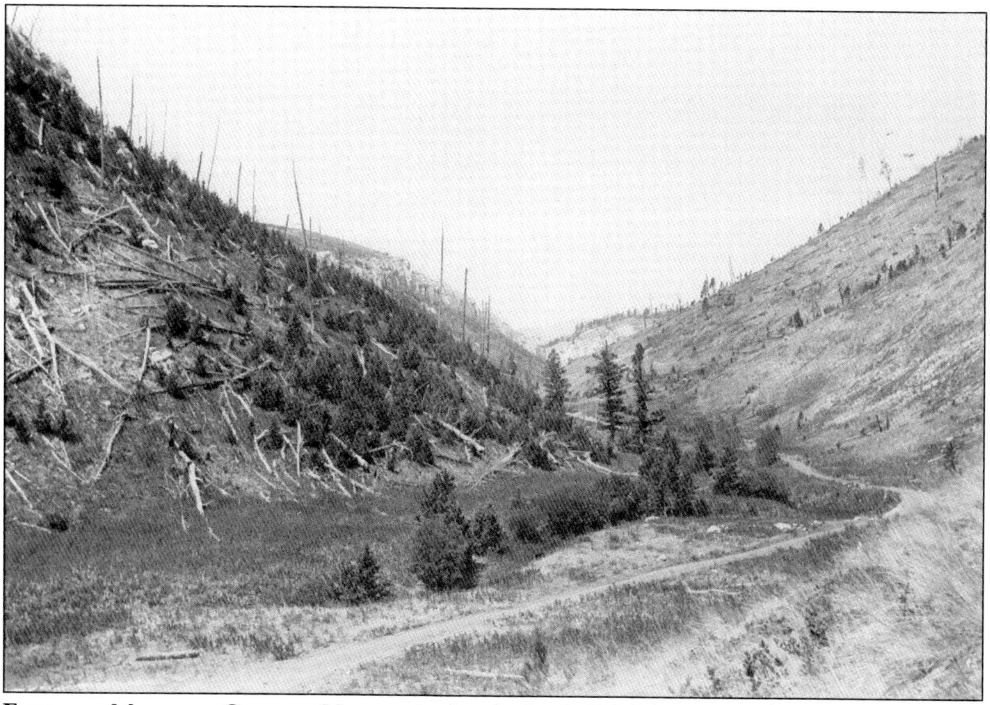

ENTERING MONARCH CANYON. Upon arrival in the Little Belt Mountains, the first display of this beauty appears as the road takes a left uphill turn at Riceville; this spot also offers an awesome view of Sluice Boxes State Park and the work done by Belt Creek. Further up the road, the scenery changes over to range and cropland until entering Monarch Canyon, just above the mountain community of Monarch. This photograph, however, does not suggest beauty so much as it does the rugged, timber-cleared side hills and a trail of seeming difficulty. (Both, MHS.)

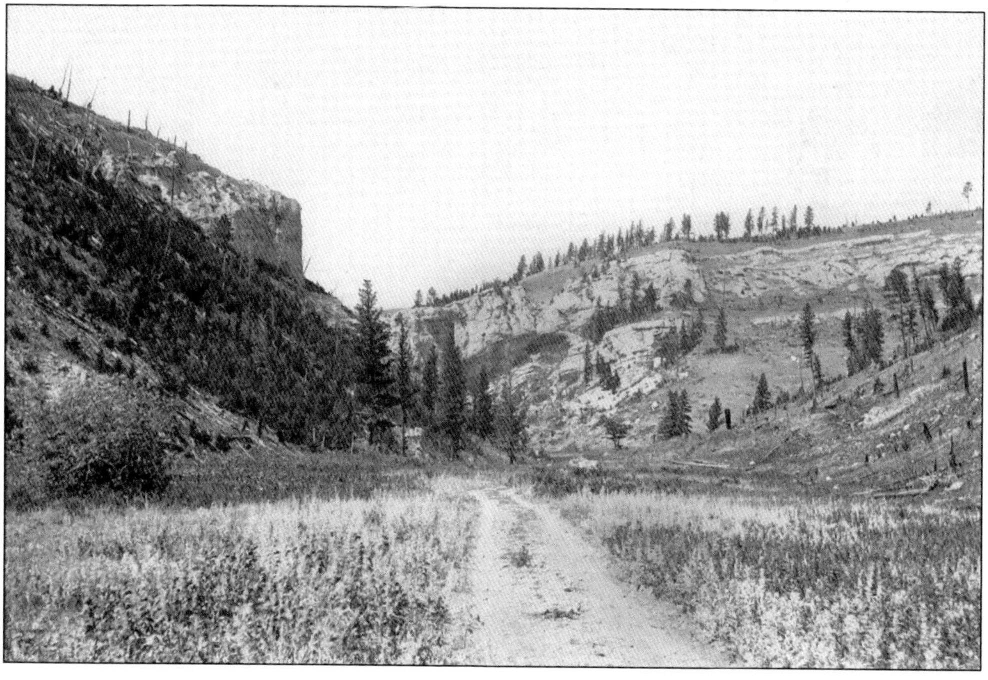

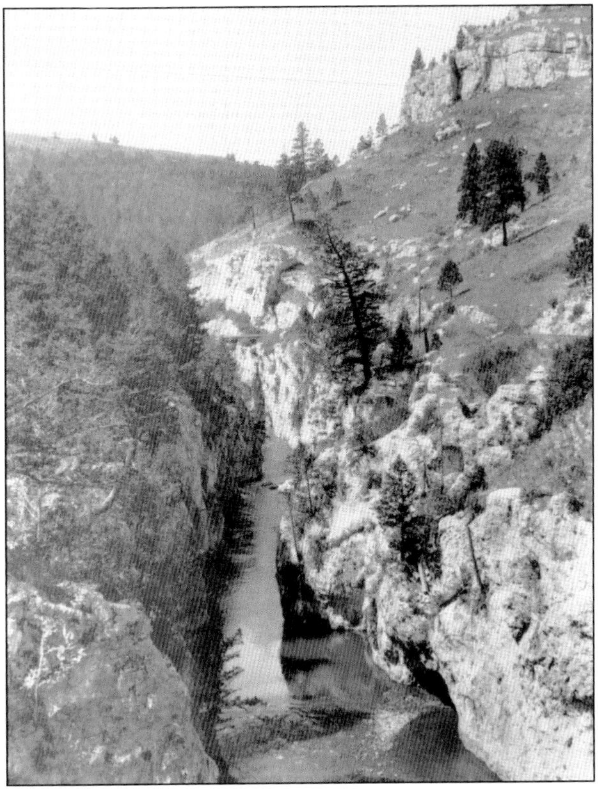

Canyon Rapids. The waters in Sluice Boxes State Park were turbulent and dangerous both for sheep grazing on both sides of the canyon and for the property owners who wanted to fish along the canyon. At the time of the construction and placement of this bridge, the area was known as the Albright Sheep Company and was owned by William H. Albright, who also owned and operated the Albright Rock Quarry. The quarry, active at the end of the 19th century, was a regular stop for the Montana Central Railroad run from Neihart to Great Falls, hauling crushed limestone to the smelter for use in processing of silver ore. (THM.)

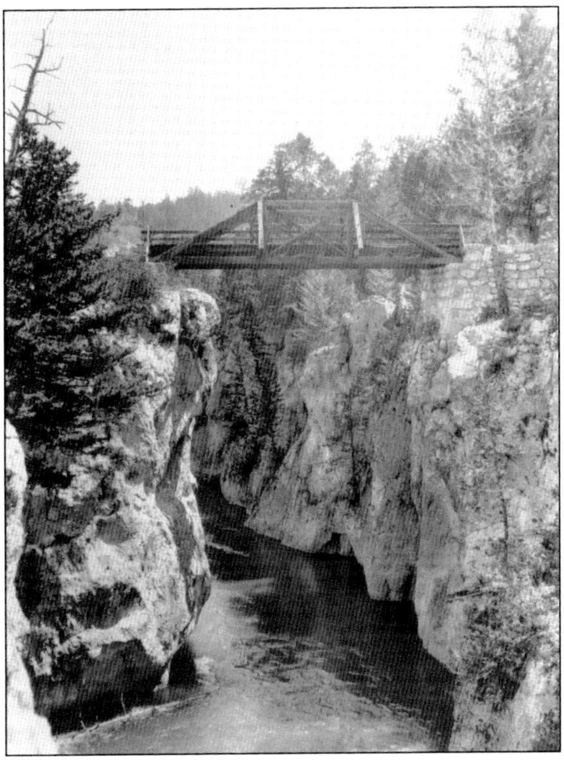

Bridge Over Turbulent Waters. In 1912, Ed Liebetrau was the manager of Albright's sheep ranch. He and his brother-in-law Frank Klepetko did a lot of hunting in the Riceville, Logging Creek, and Albright area. So much, in fact, that Klepetko ordered construction of the bridge (pictured) at the smelter in Great Falls, had it shipped out to the area on the train, and had it swung into place so that the two of them could go hunting on the other side. They also covered the bridge planking with tin so that sheep, too, could cross over to the other side. (THM.)

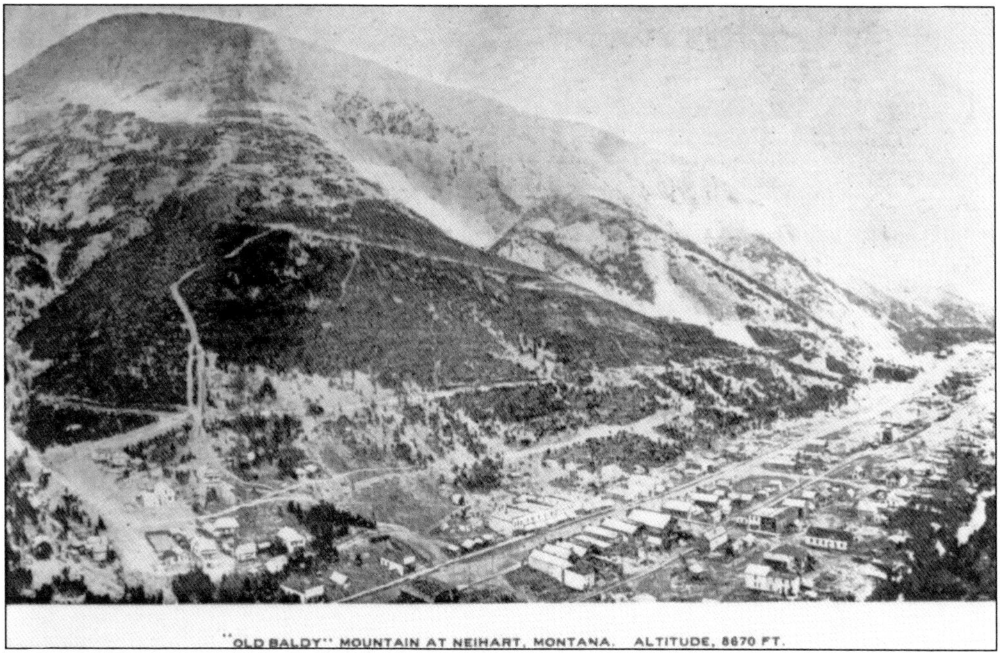

TWO PHOTOGRAPHIC CARDS OF NEIHART. "Old Baldy," as the above photograph declares, stretched the length of Neihart and this portion of the Belt Creek Canyon. The photograph was taken in the late 1890s since the railroad tracks and water tower can be spotted. Of particular interest is the pattern cut into the side of "Old Baldy," running from Rock Creek, Compromise Gulch (left in the photograph) up and across the mountainside to the large gully. Historical notes suggest this was the path of the tramway carrying ore from the Broadwater Mines to the Diamond R Concentrator. The photographic card below is a view of Carpenter Creek located on the northeastern side of "Old Baldy." (Both, THM.)

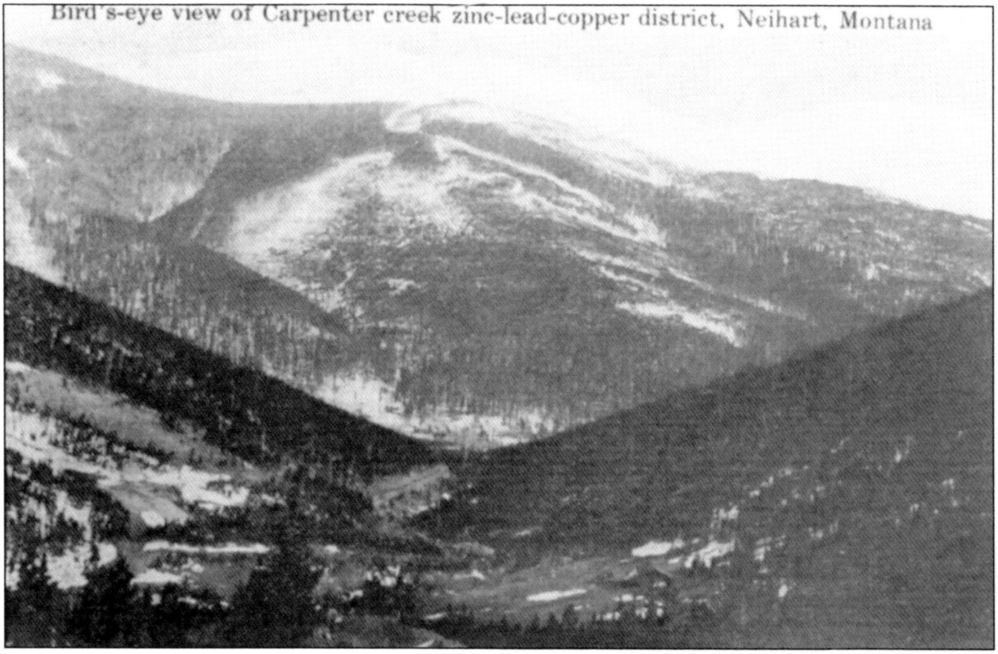

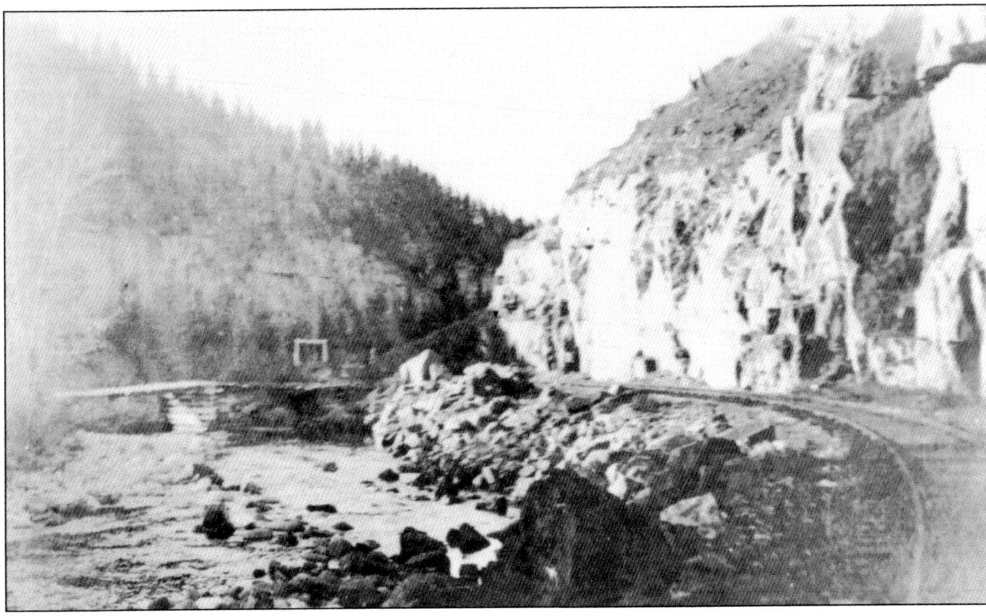

ANOTHER CURVING VIEW OF THE TRACKS AND BELT CREEK. These two photographs of the railroad tracks and bed running alongside Belt Creek clearly demonstrate that space was tight in the canyon. The bridge on the left-hand side on the above photograph is unidentified, and it is very difficult to determine its exact location. The image below, also in Neihart Canyon, is not identified other than being in the canyon between Neihart and Monarch. (Both, MHS.)

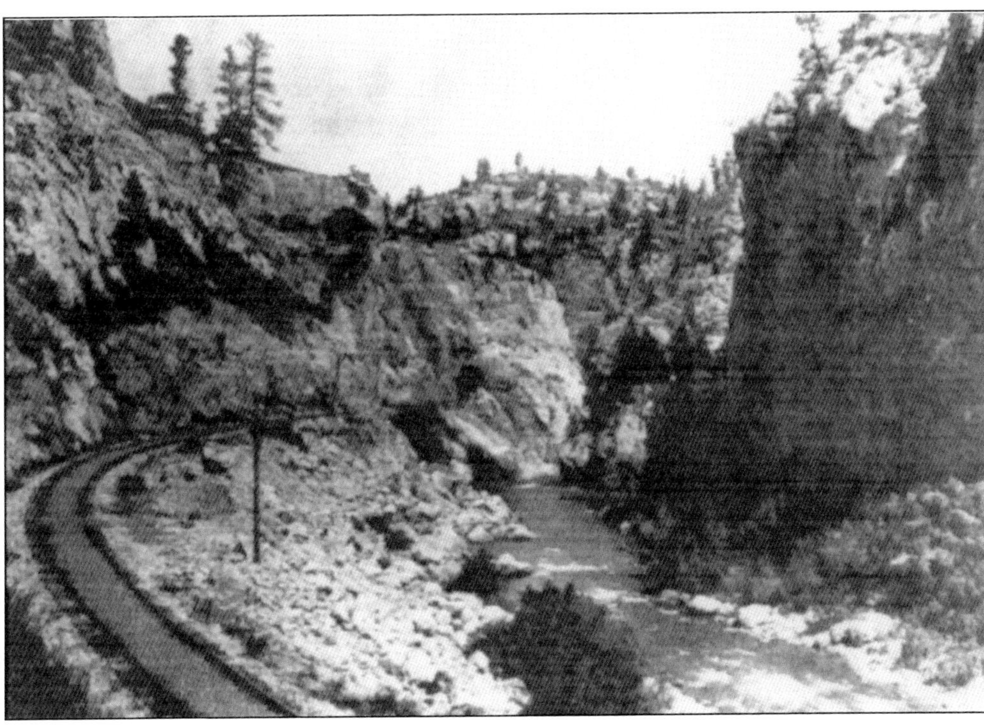

Six

Education, Religion, and Family

[Isabella Bird asked a livery stable owner if she would be safe taking a midnight ride alone. He responded:] There's a bad breed of ruffians, but the ugliest among them all won't touch you. There's nothing Western folks admire so much as pluck in a woman.

—Isabella Bird, *A Lady's Life in the Rocky Mountains*

Perhaps half of all the people who populated the early camps were involved with actual mining operations. The rest were there to serve and support those who did the mining. It did not take long for women and families to arrive in camp. Education and religion were an important part of early life in Neihart.

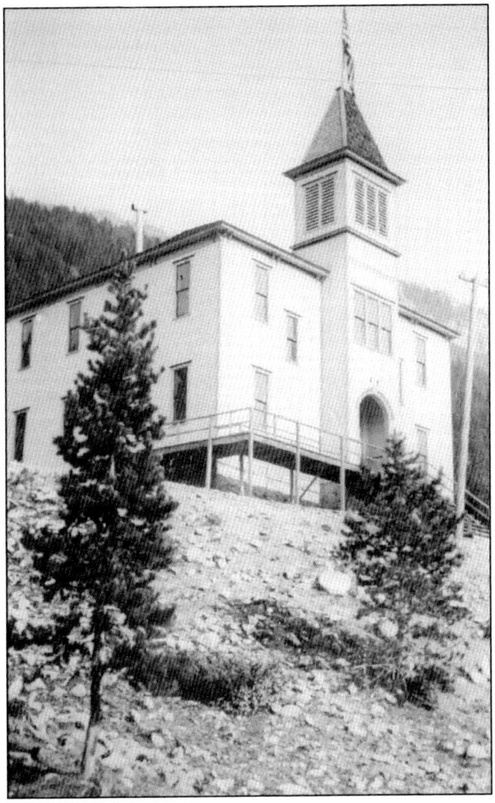

SCHOOLS AND CHURCHES. There is no doubt that faith and education were highly important to the early residents of Neihart. Along with Brother Van Orsdel and the Methodist congregation, Rev. Samuel E. Wisliard of the Presbyterian church constructed a church on Main Street and Jefferson, which is still standing today. The Episcopal church used the AOUW hall on Sundays, and the Catholic church made use of the miners' union hall for celebration of mass. Eventually, the railroad depot was converted for use as the Catholic church; it, too, is still in use. Currently, however, it serves as the frame and foundation of a newly constructed private dwelling visible on Main Street near the Neihart Inconvenience Store. (THM.)

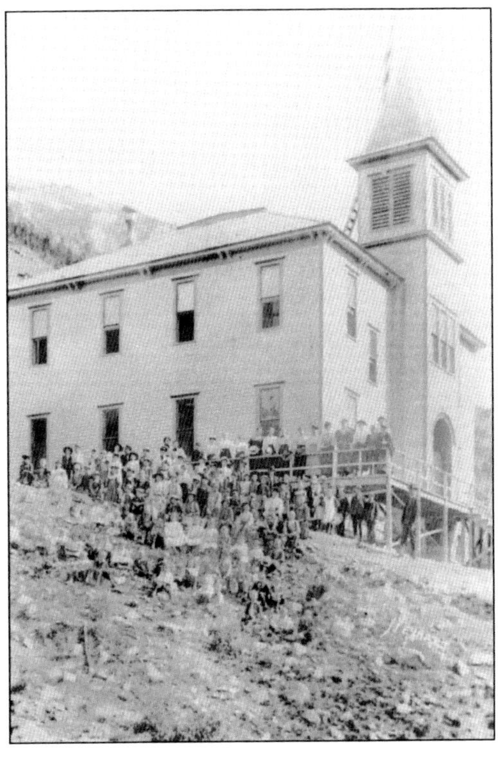

EDUCATING AREA YOUTH. The two photographs on this page, while not directly related to churches, do show that Neihart residents took their education responsibilities seriously. In 1892, the school board bonded $5,000 for seven years to construct the four-room school shown here. The new building, one block from the Methodist church on Forest and Wells Streets, served until 1940. Joseph Wagner was the contractor, and Prof. E.E. Bryant continued as principal of schools. (THM.)

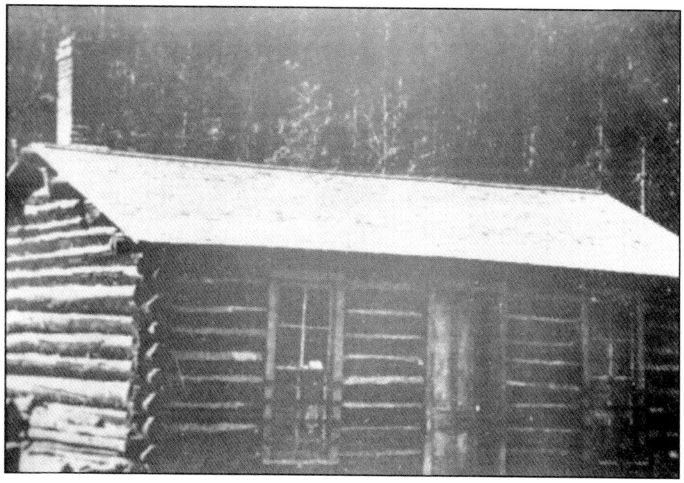

NEIHART'S HUMBLE FIRST SCHOOL, C. 1885. In Neihart's earliest days, there were not many women or families among the residents. Yet miners continued to arrive, buoyed by stories and previous experiences. Soon, their wives and children followed, creating a sudden need for a center of education. This log shack, pictured around 1885, has been identified as the first school in Neihart. (THM.)

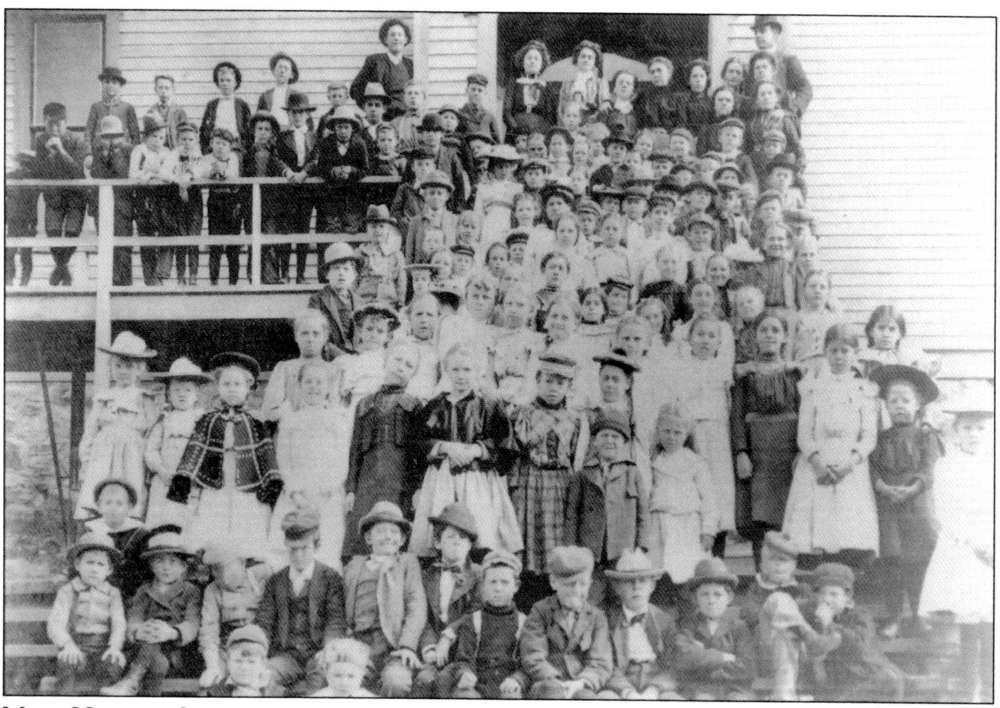

MORE NEIHART SCHOOLCHILDREN. Students pose for a photographer in front of Neihart's new school. The photograph dates to 1901, making the school about one year old. Unfortunately, the identities of the students and their instructors are unknown. Although these youngsters' parents and grandparents found the streets of Neihart paved in sweat and dirt rather than silver and gold, they understood that the first step to success was a quality education, and they were determined to provide that to their children. (THM.)

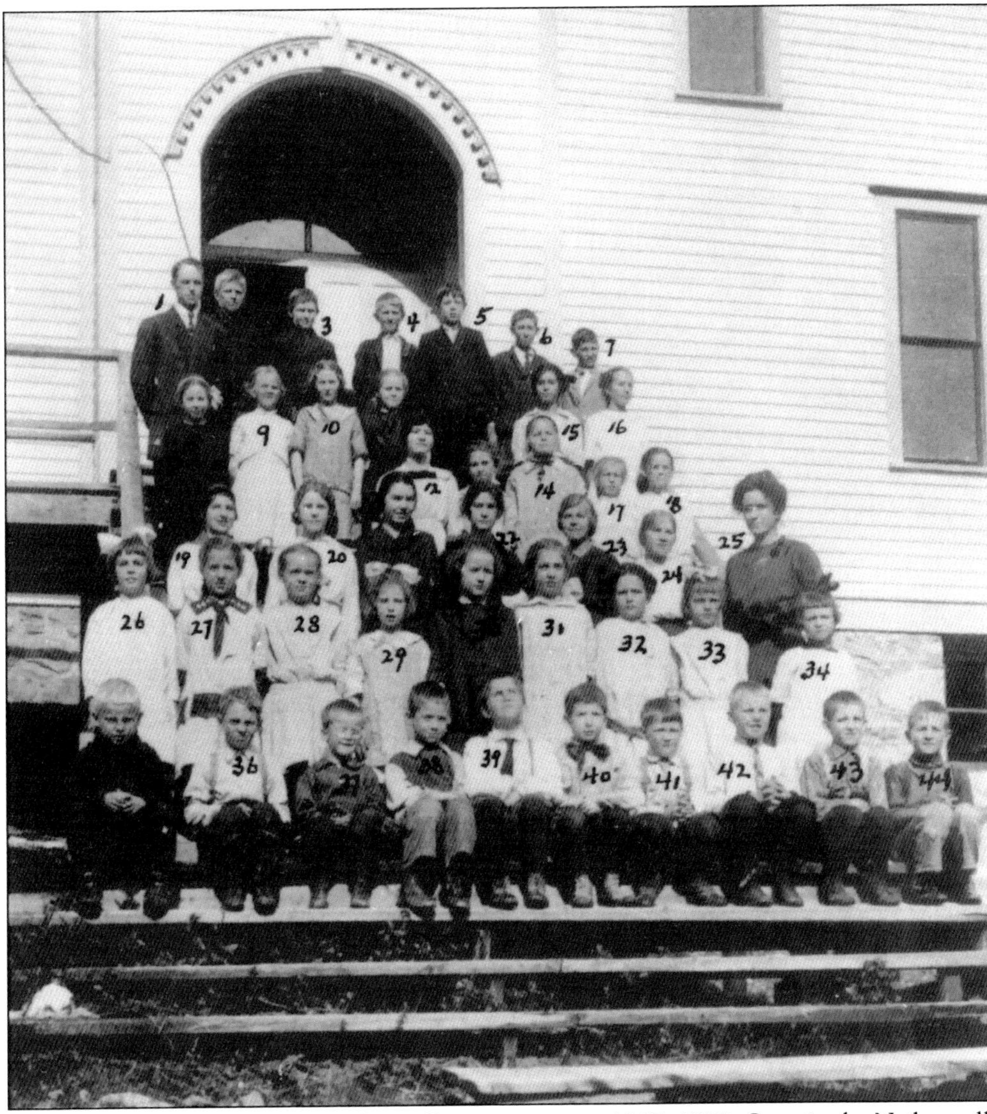

NEIHART'S SECOND SCHOOL ALL-CLASS PHOTOGRAPH, C. 1915–1920. Seen in the Neihart all-school picture are, from left to right, (first row) Charlie Swanson (No. 35), Jack O'Connor (No. 36), Oscar Shoberg (No. 37), Mauritz Shoberg (No. 38), Danny Lenny (No. 39), Danny Redden (No. 40), Carl Tripp (No. 41), Blake Westgard (No. 42), Charlie Darnell (No. 43), and Ed "Buster" Meager (No. 34); (second row) Eloise Barker (No. 26), Gladie Ledbetter (No. 27), Impi Gavander (No. 28), Sweet Stark (No. 29), Dorothy Powers (No. 30), Florence Barker (No. 31), Kathryn Bronson (No. 32), Agnes Gavander (No. 33), and Gil Meagher (No. 34); (third row) Stella Ledbetter (No. 19), Dorcas Shaw (No. 20), Margaret Lenny (No. 21), Rose Lenny (No. 22), Sadie Seeger (No. 23), Emma Anderson (No. 24), and Lillian Powers (teacher, No. 25); (fourth row) Olivia Oriel (No. 12), Isabel Forcum (No. 13), Harriett Westgard (No. 14), Mamie Seeger (No. 17), and Bessie Darnell (No. 18); (fifth row) Adla Shoberg (No. 8), Myrtle Swanson (No. 9), Evelyn Redden (No 10), Alta Shoberg (No. 11), Kathleen Lenny (No. 15), and Edna Ledbetter (No. 16); (sixth row) teacher J.B. Walsh (No. 1), Raymond Swanson (No. 2), Teddy Morrison (No. 3), Reuben Swanson (No. 4), Bun Stark (No. 5), Glen Darnell (No. 6), and Carl Faller (No. 7). (THM.)

METHODIST "BROTHER" VISITS CAMP. "Brother Van," as the Reverend Van Orsdel was known to his many parishioners, arrived in Montana in the summer of 1872. He conducted the first organized church service in August 1882. This service was held in a log cabin belonging to Joseph Hannah on a site eventually occupied by Cook's Furniture and Undertaker Store in the mid-1890s. From the time of his arrival until his death in 1919, Brother Van was instrumental in establishing more than 100 churches and parsonages throughout Montana, as well as the Montana Wesleyan University and the Montana Deaconess School for Children, both located in the Helena Valley, southwest of Neihart. (THM.)

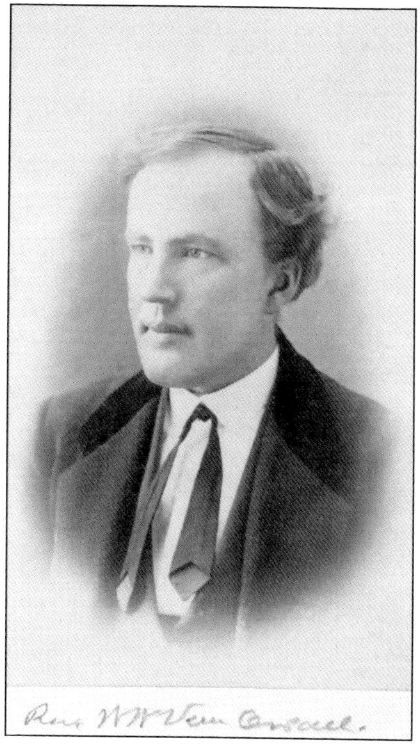

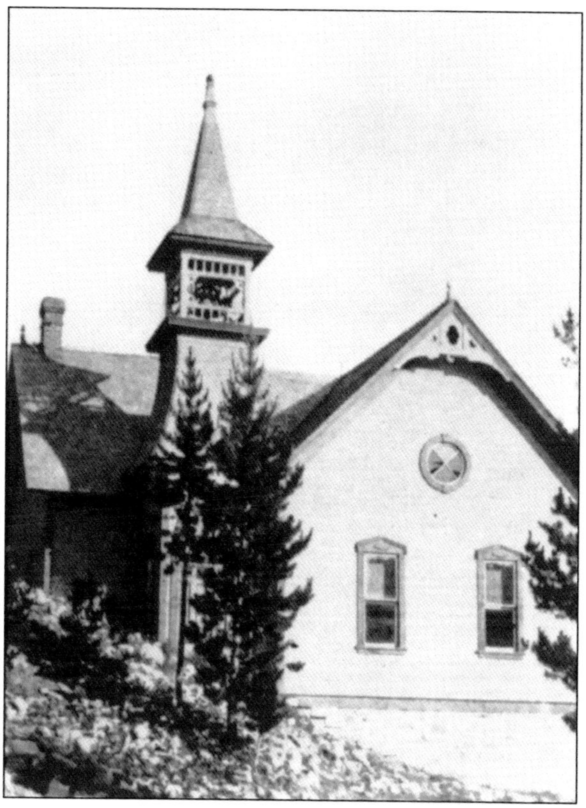

METHODIST EPISCOPAL CHURCH OF NEIHART, C. 1895. Standing on the southwest corner of Forest and Hulbert Streets, south of the four-room school, the First Methodist Episcopal Church commanded a beautiful view and projected a magnificent image up and down the town. Situated apart from the daily noise and doings of the Neihart streets, it was convenient for most parishioners anytime they needed. The church was constructed in 1892, while Rev. Joel Vigus was pastor. In 1895, Reverend Noble became pastor of the 50-member congregation. (THM.)

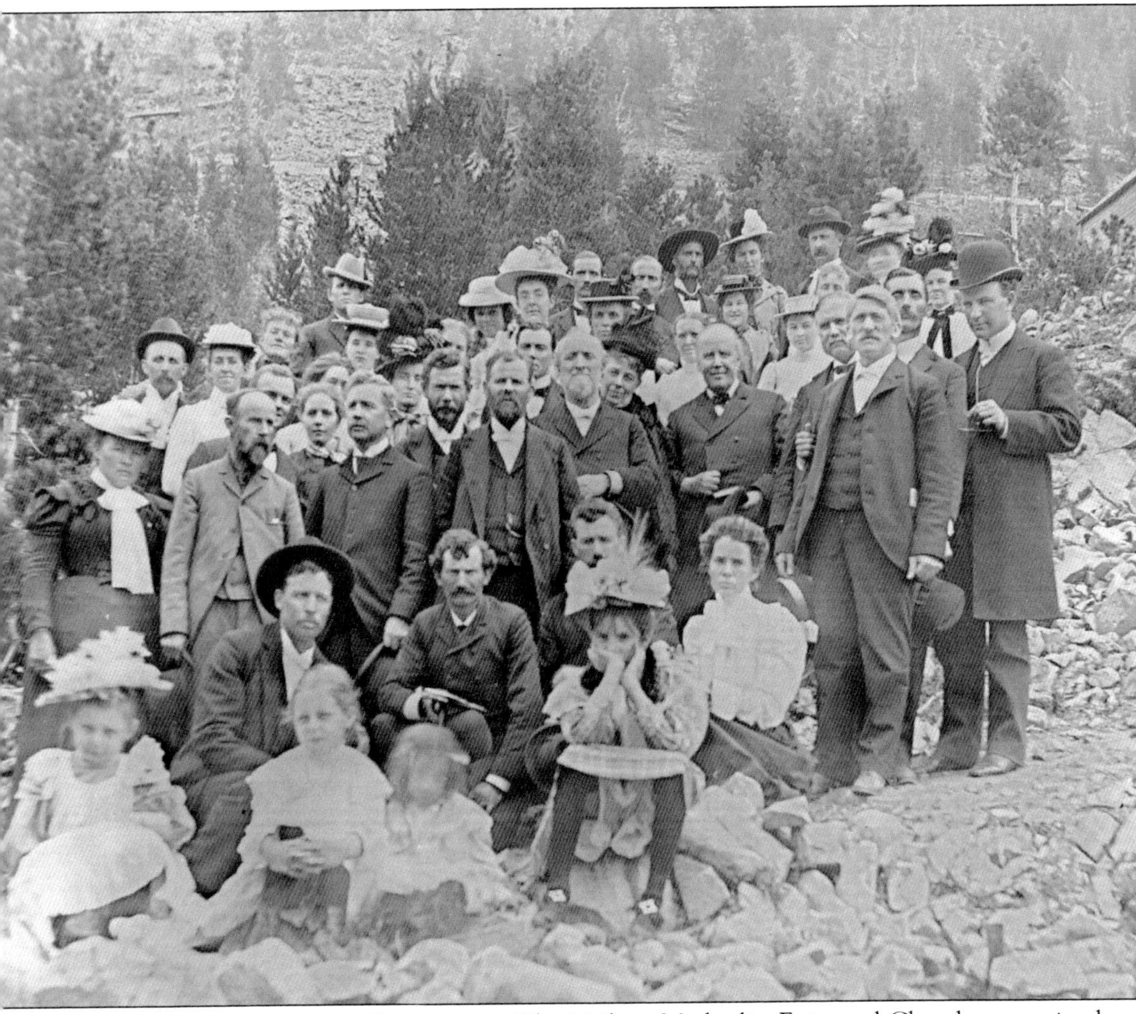

Hosting the Annual Conference. The Neihart Methodist Episcopal Church entertained the state delegation at an annual gathering; it is uncertain exactly what year this photograph was taken. The Reverend Van Orsdel, "Brother Van," standing fifth from the right, smiling, and holding a book in his left hand, held two early meetings in Neihart in August 1892 in a lone log cabin. Of that first meeting, Donald Mackintosh, Neihart's first postmaster and town recorder, unabashedly declared that he had "attended two . . . very impressive religious services . . . the first in the middle of the Atlantic Ocean . . . surrounded by gale winds and seas terrible . . . and [the second] at Neihart—when this mountain glen first echoed man's songs of prayer and praise to the great Creator." The second service he spoke of was offered by Brother Van. It was not long before Neihart congregants built their own church to honor both Brother Van and their Lord. The new church was nestled up the east slope of Long Baldy Mountain, on Forest and Hulbert Streets. (MHS.)

MYRTLE E. NEIHART. Myrtle, born February 25, 1891, was the only child of James and Sarah Neihart. Both Neihart newspapers, the *Herald* and the *Miner*, kept tabs on the town's first family, reporting their comings and goings. Events such as their Yellowstone vacation, Myrtle's attending school in Helena, and visits from her stepsisters Ida and Mae were documented. Myrtle seemed to have had a typical mountain community life, enjoying the great outdoors with friends, picnicking and hiking the mountain trails. Myrtle died in Youngtown, Arizona, on December 31, 1971, at 80 years of age. (Carol Zehntner.)

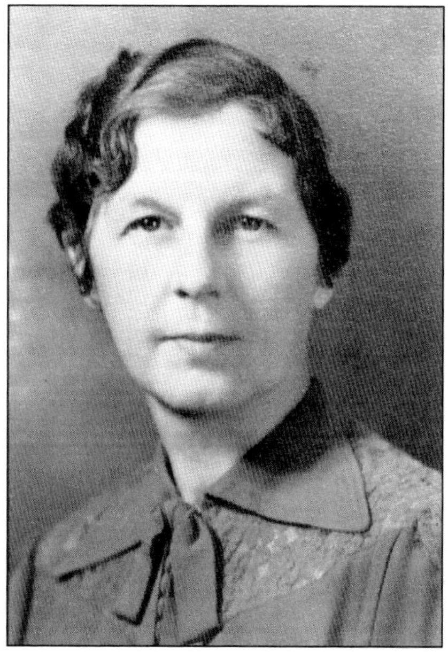

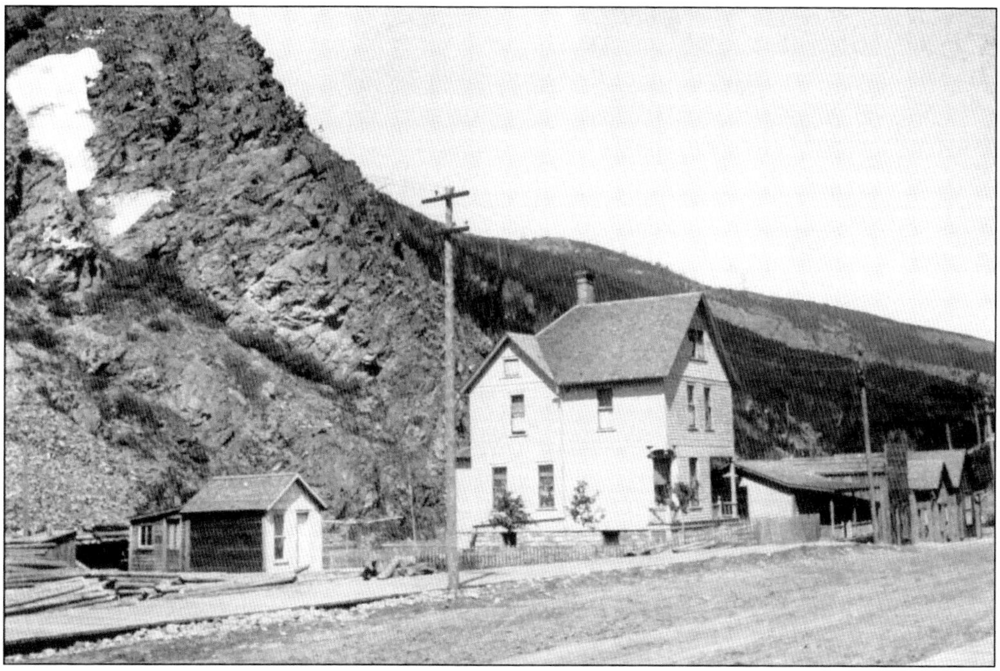

THE NEIHARTS' FIRST HOUSE TOGETHER. James Neihart built his family's first home across the street from his mine, the Queen of the Hills, and the Frisco Hotel. James L. Neihart and Sarah E. Patrick Sutton first met when they both resided in Fort Benton around 1879 to 1881. They each followed prospectors to Barker-Hughesville, where their friendship grew. Sarah was married to Solomon H. Sutton and later divorced. In 1889, Sarah and James married, and with a household of four children—daughter Myrtle and (from Sarah's first marriage) Ida E. Sutton, born January 1875; Mae E. Sutton, born May 1877; and Henry W. Sutton, born June 1879—the family needed a large home. They lived in this house until James sold the Queen of the Hills along with the home. (THM.)

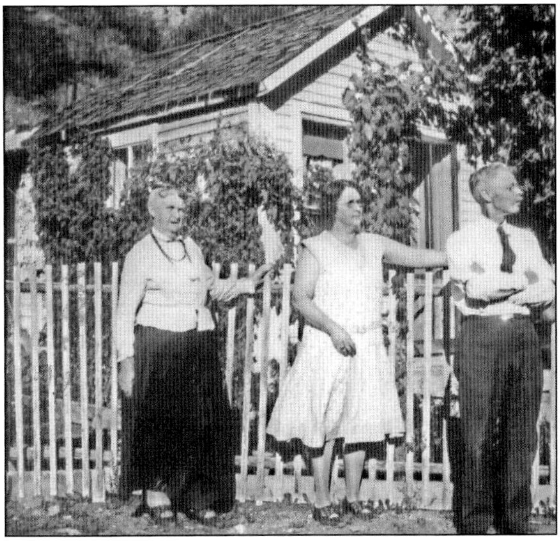

Sarah Neihart Poses in Front of the Cabin. Sarah Neihart (far left) and an unidentified couple pose in front of "the Cabin," as it was affectionately known. Sarah spent the rest her life in Neihart, having previously lived in California, Oregon, Utah, and, within Montana, High Plains, Fort Benton, and Barker. She was born on February 27, 1857, to Nancy A. Morton Patrick and Col. J.W. Patrick near Crescent City California. When only 10, Sarah's mother lost her eyesight due to snow blindness, leaving Sarah to take over the household duties. (Carol Zehntner.)

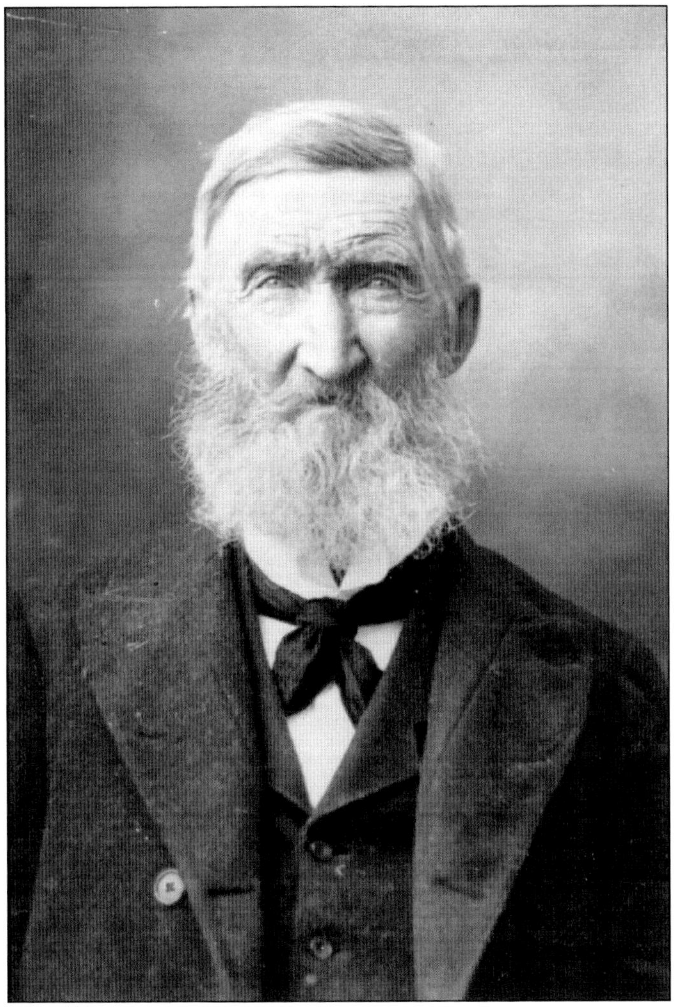

J.W. Patrick. Sarah Neihart's father, John Walker Patrick, is one of the most interesting of Montana's pioneers. Born near Boone Licks, Kentucky, on August 17, 1809, Patrick was the son of William Jefferson and Jemima (Walker) Patrick. William died during the War of 1812, leaving Jemima and three sons. John Walker Patrick was a schoolmate and good friend of Abraham Lincoln. He studied medicine for two years and moved to St. Louis at age 18, in 1827. He married Frances X. Coffey, and the couple had three girls. When Frances died, Patrick left the girls with her family and started up the Missouri River with 20 men, reaching Montana on September 3, 1836. As a scout, trader, and wagon boss, he lived in several other states. He remarried and eventually he and his family, including Sarah, made it back to Montana. (Carol Zehntner.)

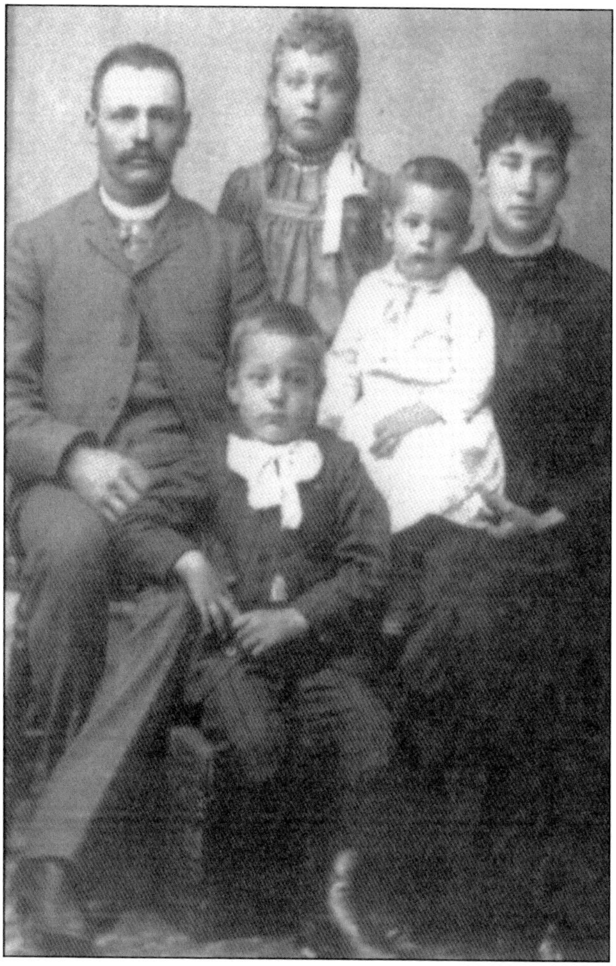

RICHARD BENNETT'S FAMILY AND HOME." Neihart's biggest booster was Richard Bennett. Born in Ontario, Canada, on December 4, 1851, Bennett was raised on a farm. He spent time clerking in a local store and practiced land law in Grand Forks, North Dakota, in 1878. He was a delegate to the North Dakota Constitutional Convention in Bismarck and also a probate judge. Bennett married Louise Doyle, also of Renfrew, and they arrived in Neihart on June 10, 1891, opening a law office where he practiced for many years. Bennett was city attorney, a position to which he was appointed upon the incorporation of Neihart. According to an 1895 edition of the *Neihart Herald*, "He speaks his mind, allies himself with the best interests of the town, and is unanimously respected." The Bennetts had five children: Mary, Edward, Anna, Richard, and Helen, but it is uncertain which three are pictured here. The location of this house is unidentified. (Both, Richard Bennett.)

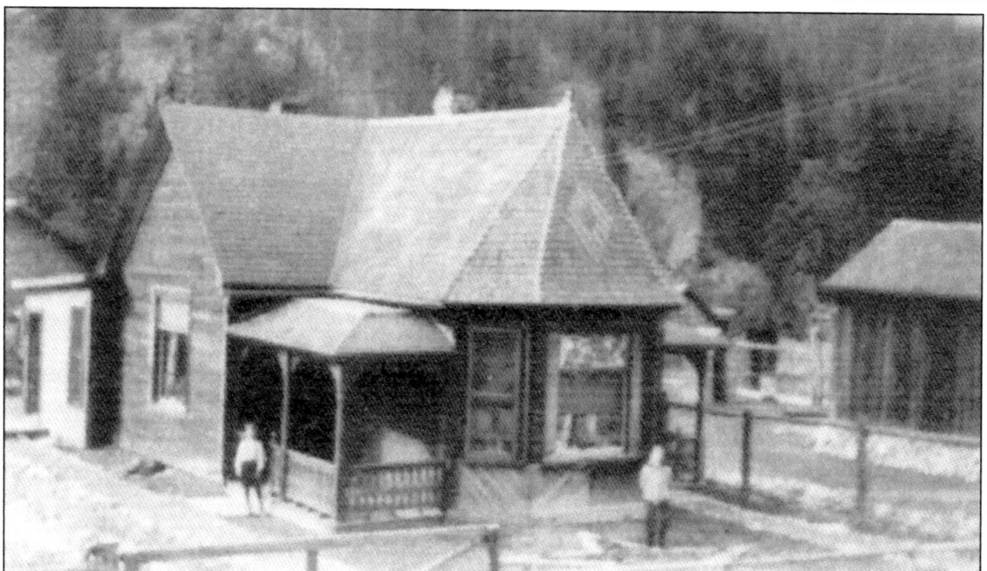

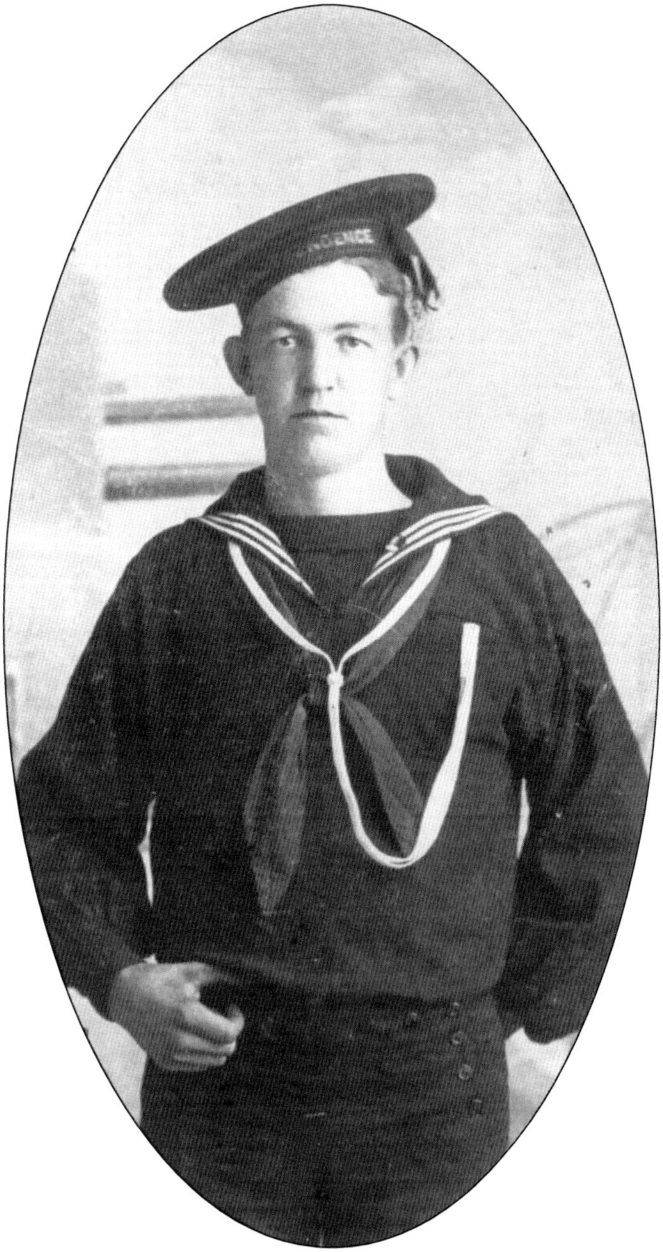

Henry W. Sutton. Henry W. Sutton was born in June 1879 in Fort Benton, Montana Territory, the third child and only son of Solomon H. and Sarah (née Patrick) Sutton. His parents had moved to Fort Benton from Missouri in 1876, along with his grandfather's three grown daughters and his two older sisters, Ida and Mae. Henry married Myrtle K. Collette of Chippewa Falls, Wisconsin, shortly before enlisting in the US Navy around 1913. He mustered out about five years later. Henry and Myrtle had three children—Fay C. (born around 1917 and later married to Ellis Julian), Eloise J. (born in 1924 and later married to David Cummings), and Wallace C. (born around 1915). Henry was employed by the *Neihart Herald* for a number of years, and he and Myrtle lived in Neihart for 50 years before moving to Great Falls around 1954 or 1955. Henry died in 1955, and Myrtle followed two years later. (Carol Zehntner.)

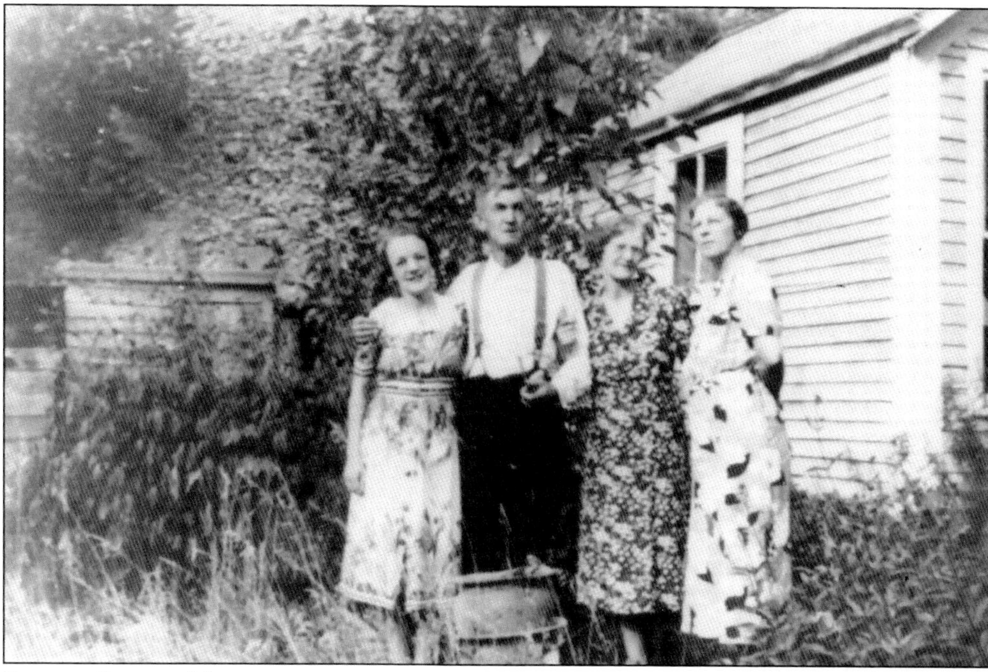

SUTTON-PATRICK MARRIAGE. Pictured from left to right are Doris Reisinger, the daughter of Ida Sutton Reisinger; Henry W. Sutton; Mae Sutton Larson (J. Leonard Larson); and Myrtle Neihart Woodward, the only child of James and Sarah Neihart. They stand beside "the Cabin" as Sarah referred to her and James's second house located on the north end of Neihart. The siblings, usually as one family, would make trips with the Neiharts, such as their grand excursion to Yellowstone National Park in 1892. (Carol Zehntner.)

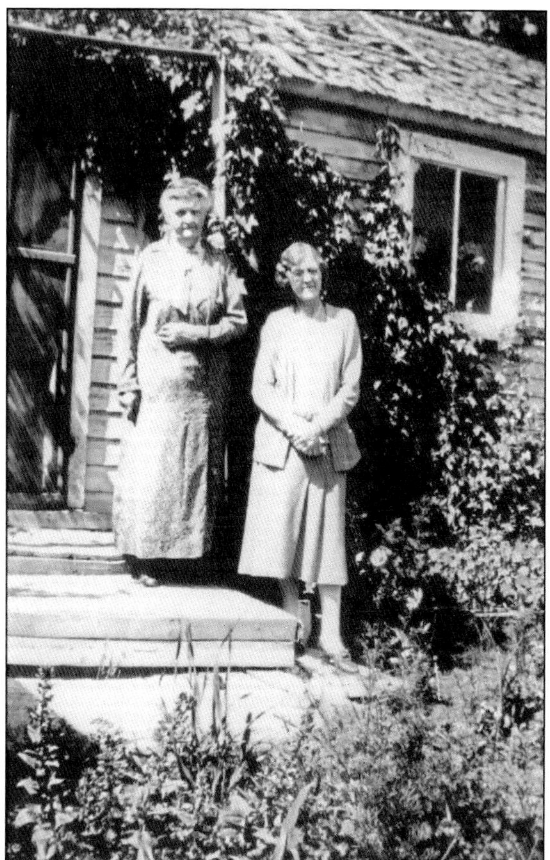

NEIHART-WOODWARD MARRIAGE. This image, also located at "the Cabin," features Sarah Neihart (left) and her daughter Myrtle. Sarah's only child with James Neihart, Myrtle married Christian Walter Woodward on January 16, 1912, and moved to California soon after. The Woodwards had two sons, Ken and Phil, but no grandchildren. According to family members, Phil would have been the last direct male descendant of James L. Neihart. (Carol Zehntner.)

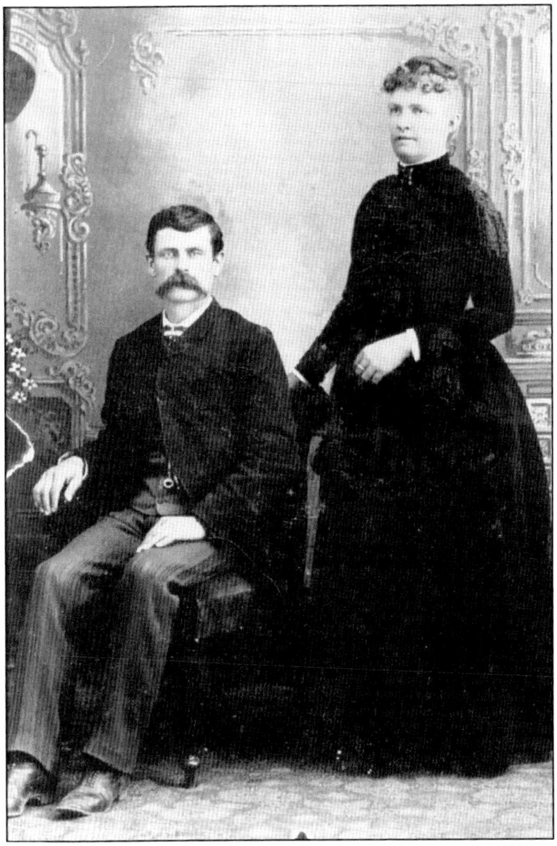

The Newlyweds. James L. Neihart and Sarah Parker Sutton first met in the mining camp of Barker in Montana Territory between 1879 and 1881. Neihart was prospecting from Fort Benton where Sarah also lived with her husband, Solomon Sutton, and their three children. At some point, both Neihart and the Suttons moved to Barker hoping life would take a turn for the better. During that time period, the Suttons were divorced with James and Sarah striking up a relationship that led to their marriage 10 years after Neihart had hit his pay load—the Queen of the Hills. (Carol Zehntner.)

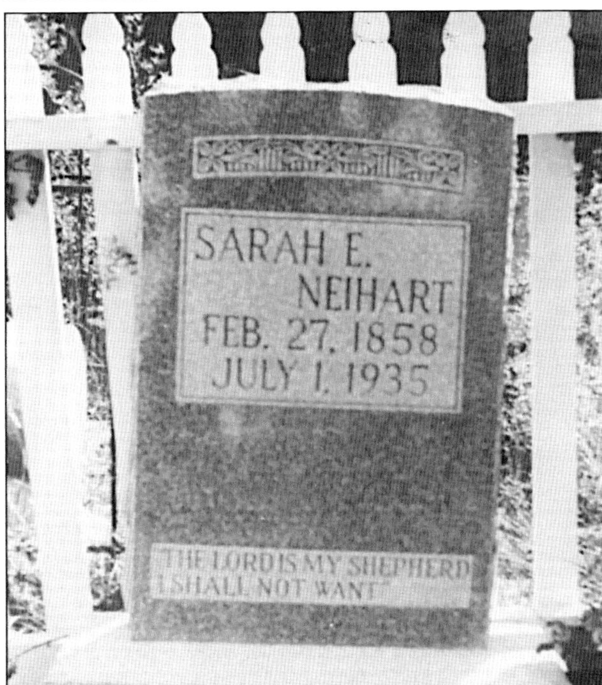

Sarah's Resting Place. There is no denying that Sarah Parker Solomon Neihart lived a full, productive, and gracious life. After James passed away, she remained a resident of the community named for its first mayor and founder—and her husband—James L. Neihart. She found peace in these hills, and in these hills is where she chooses to stay. (Author's collection.)

Seven

RECREATION

Mr. President, I want to show you the best exhibit I ever saw, Montana.

—James J. Hill to Pres. William Howard Taft, 1909

In his *Geology Study of the Little Belt Mountains, Montana* in 1889, W.H. Weed wrote of the approach toward the range from the west, north, or south:

> Here the abrupt transition from soft to harder rocks is strikingly shown in the resulting scenery. Streams in the range pass through narrow clefts in the massive limestone. Along the larger streams (Smith, Hound, and Belt) such gorges are wider, though generally impassable, with the limestone presenting a bewildering array of pinnacles and towers, castle-like masses, natural archways, and ribboned walls.

See what the pioneers saw, what they did, and how they did it in order to enjoy the beautiful scenery all around them.

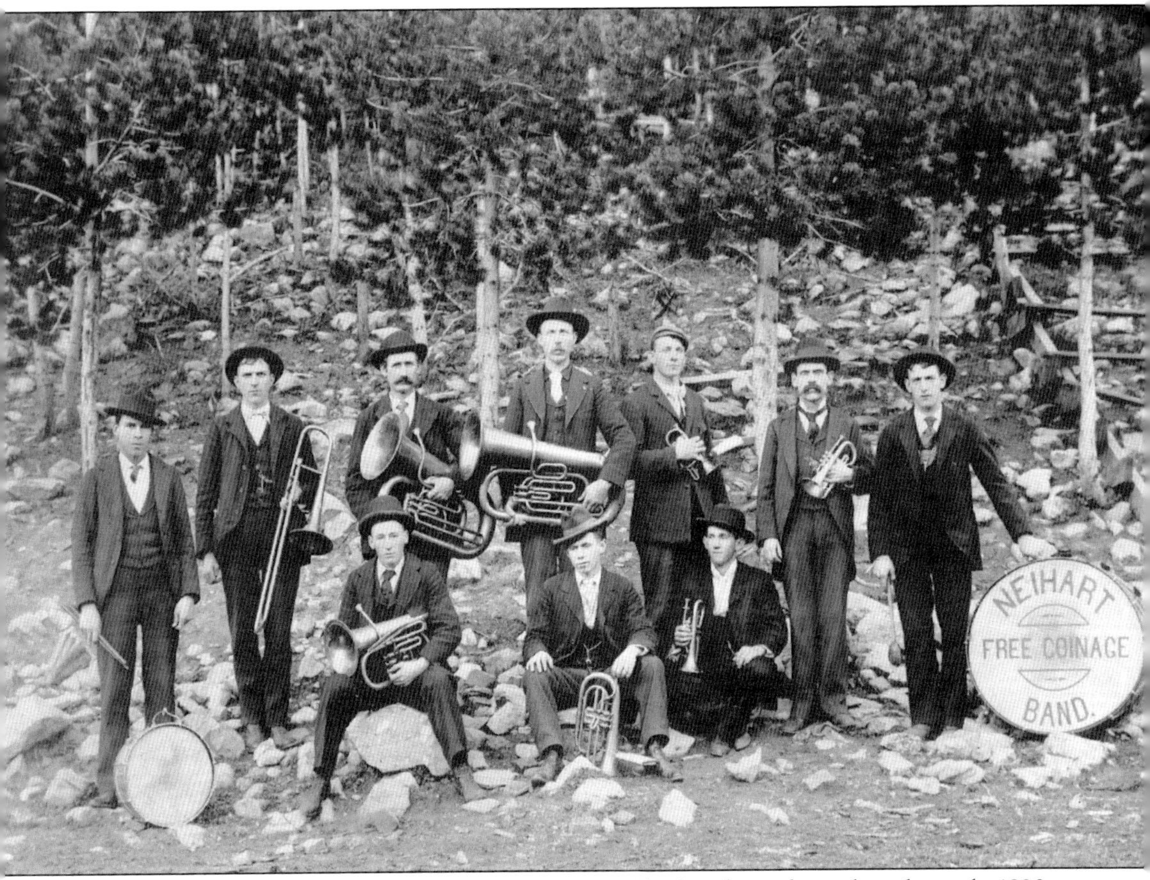

Neihart Free Coinage Band, October 1890. The band was formed in the early 1890s as a protest of the Sherman Silver Purchase Act of 1890 and the movement to use gold exclusively to back the US currency. Naturally, silver mine owners and miners wanted the monetary system to use both gold and silver, or silver exclusively. They achieved neither, but the Neihart Free Coinage Band did have a degree of success. Band members were, from left to right, (first row) two unidentified musicians and Dave (or Darel) Marelius (cornet); (second row) unidentified, Carl Marelius (trombone), three unidentified musicians, ? Abbot (cornet), and Charles Bonner (bass drum). (MHS.)

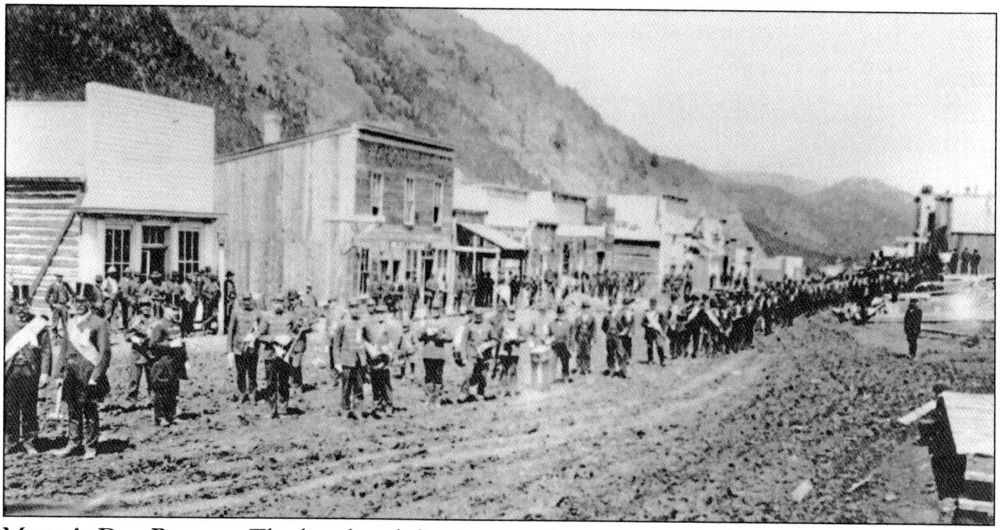

Miner's Day Parade. The band and dignitaries taking part in Neihart's Miner's Day parade did not find their Main Street paved with silver or gold but instead had to navigate through the wagon-wheel ruts; evidently, the town had not yet completed construction of the boardwalk. Members of the marching band are unidentified, but may include members of the famous Neihart Silver Coinage Band. (THM.)

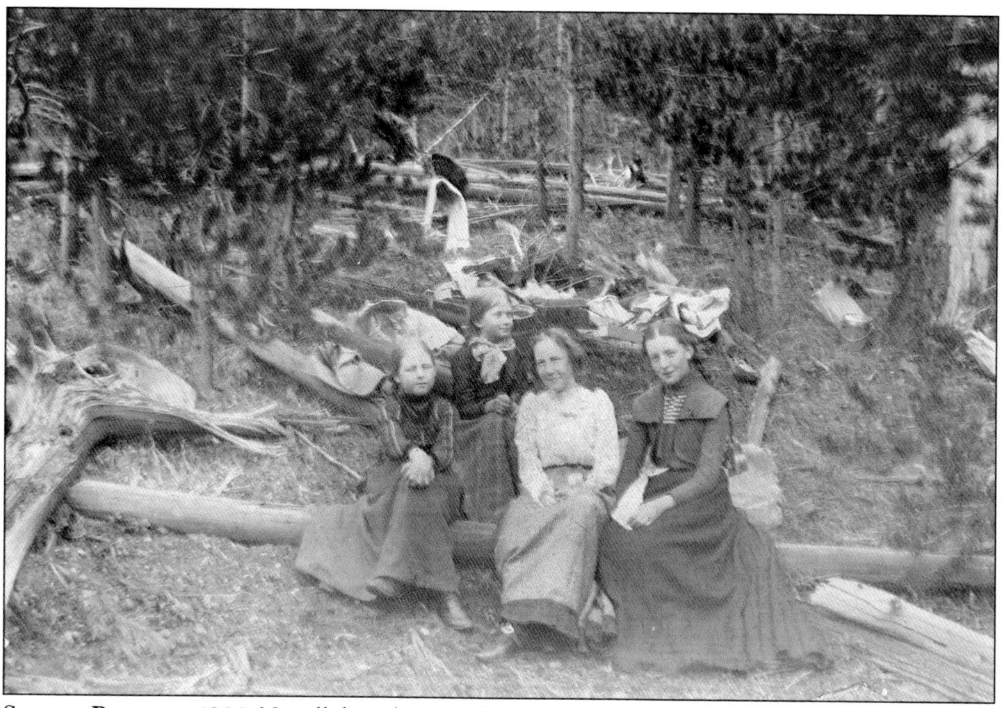

Summer Picnic, c. 1900. Not all those living in Neihart spent their days underground. Photographed around 1900 are, from left to right, Myrtle Neihart (age nine), Alice Harrison (age nine), Isabel Stone (age unknown), and Frances McCool (about 12) enjoy the fresh mountain air at the Neihart picnic grounds above the town, near Belt Creek. Myrtle was the only child of James and Sarah Neihart, Alice Harrison was the daughter of William and Sarah Harrison, and Frances McCool was the daughter of Edward McCool. (MHS.)

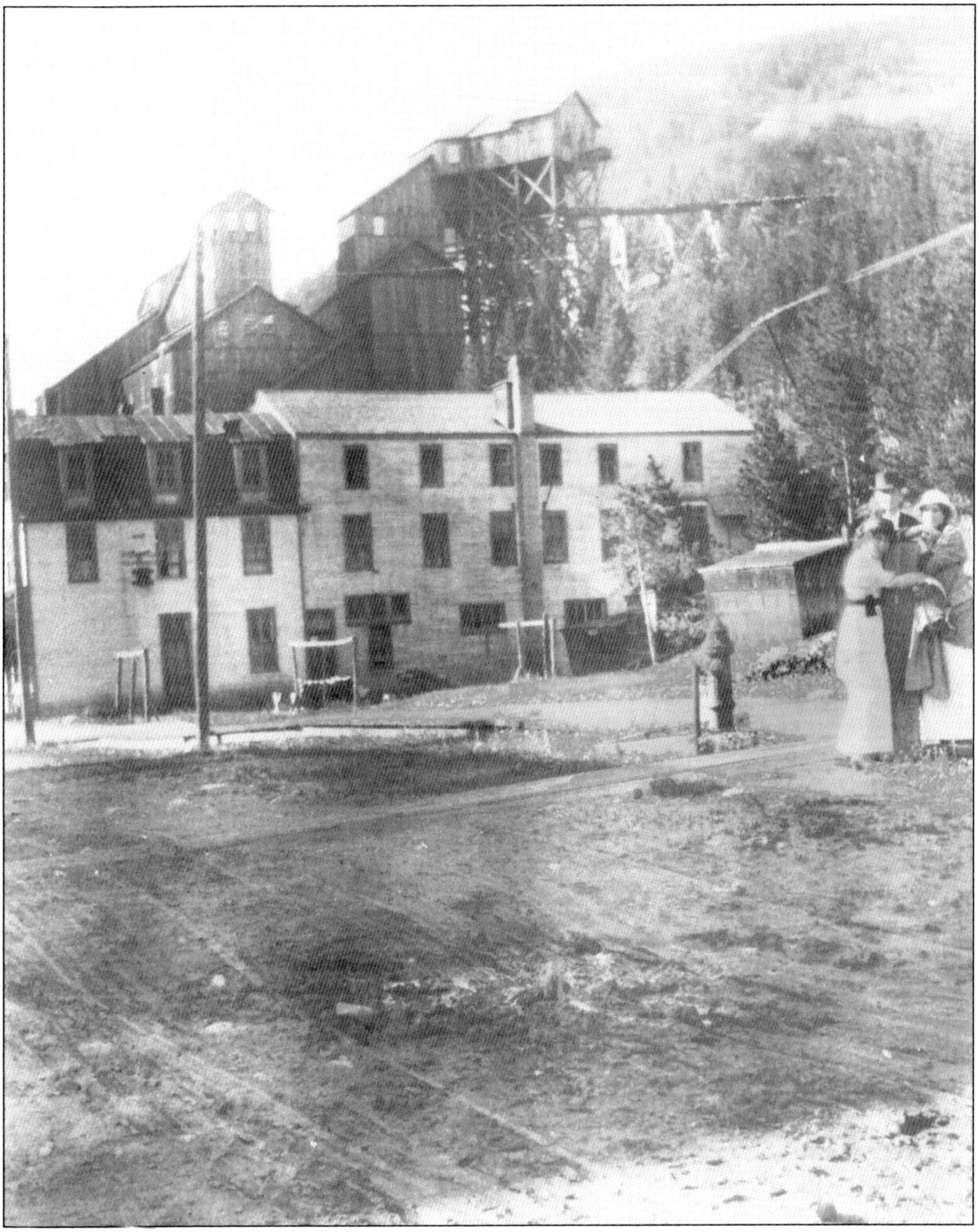

MOTHER, FATHER, AND DAUGHTER ENJOYING A STAY AT THE FRISCO. What differentiates this view of the Frisco Hotel is the view of Neihart's skyline around 1901. It is difficult not to notice the Diamond R Concentrator above the left corner of the Frisco. Even harder to miss is the mining works of the Queen of the Hills. Because of its elevation, it towers over everything in Neihart. The fire hydrant just to the left of the three posing individuals provides evidence of the Neihart Water Company; established in 1892 (at a cost of $40,000), its initial stated purpose was to provide the community with water for fire protection. A reservoir was built one-and-a-quarter miles above Neihart, and 25 dry hydrants were installed, along with a 10,000-foot-long, 12-inch water main connecting the fire hydrants. A natural fall of 300 feet provided an estimated water pressure of 150 pounds per square inch. (MHS.)

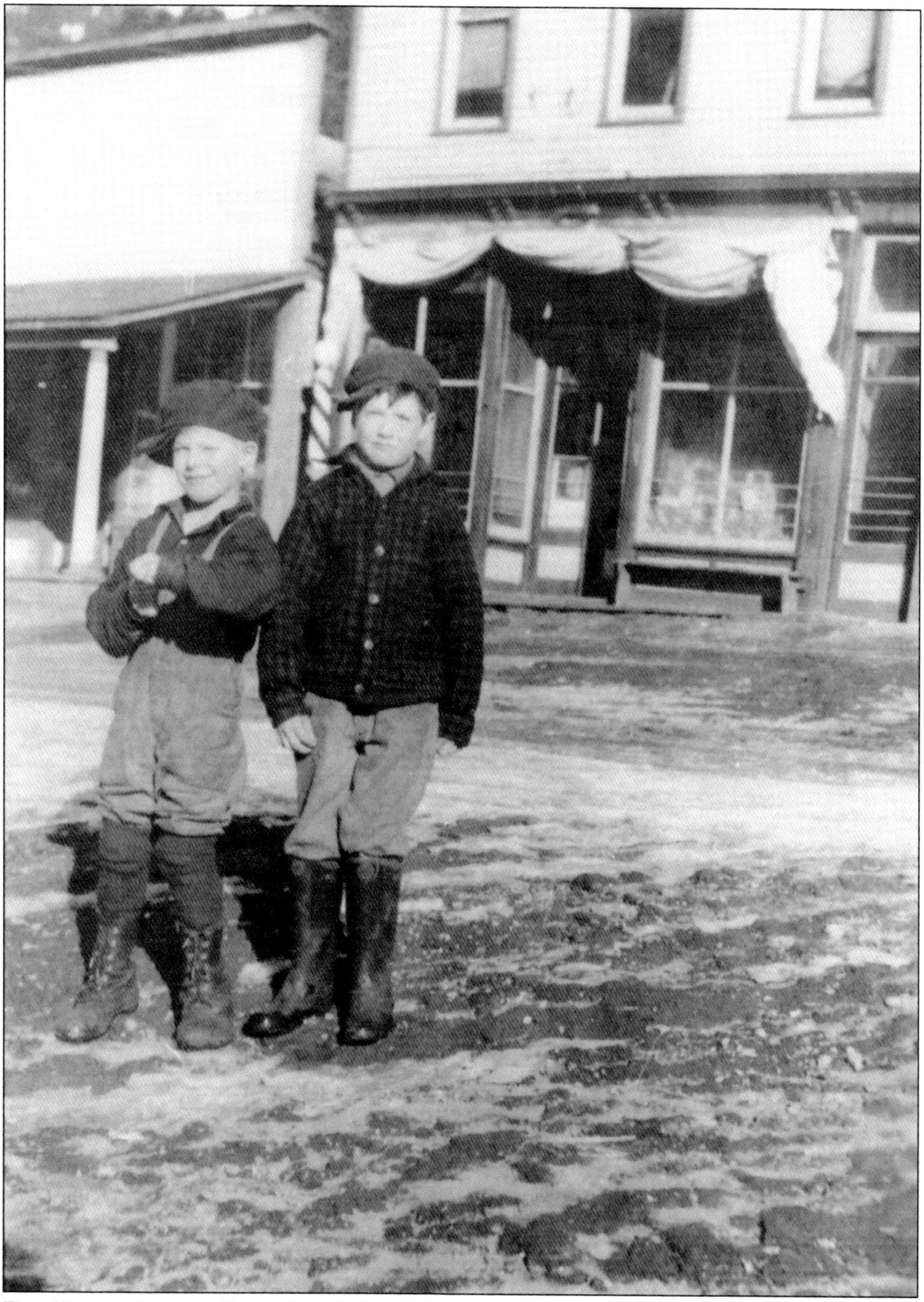

Two Young Neihart Lads. Joe Taylor (left, about eight years old) and his buddy Howard G. O'Conner (around the same age) stand in front of a barbershop, dressed warmly in hats and boots. It looks as if there may have been some snow on Neihart's Main Street. (THM.)

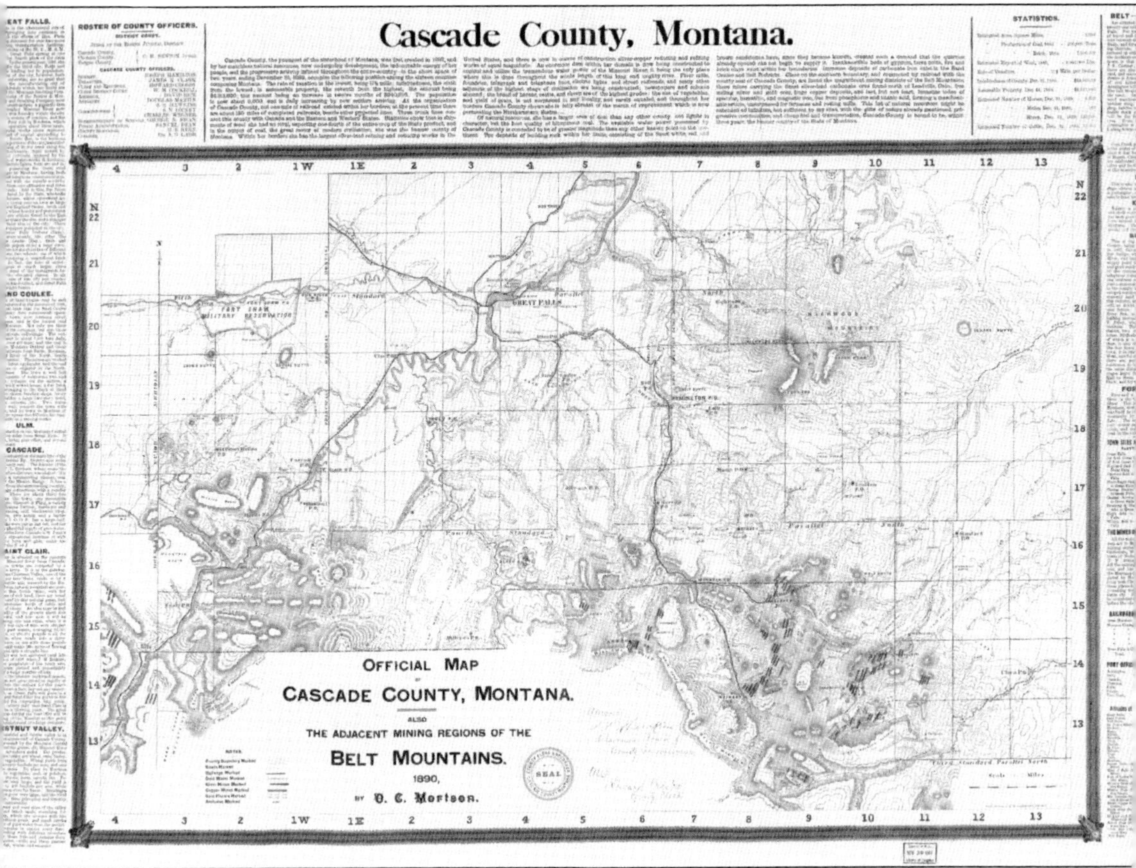

O.C. Mortson's 1890 Cascade County Map. Oswald C. Mortson is a very interesting character; he arrived at Fort Shaw with Col. John Gibbon in 1870 when the 7th US Infantry relieved the 13th Infantry. He was a sergeant in Company C. He was also a self-taught geologist, cartographer, and author. He had a bit of an artistic flair, liking to "fancy up" some of his maps, particularly his 1893 map of Great Falls and the surrounding area. After his tour of duty, Mortson mined for a while before eventually serving as the Cascade County surveyor. Some of his early mapping for the county has been included because of its distinctiveness and accuracy; this 1890 map includes brief biographies of the communities in the county. (THM.)

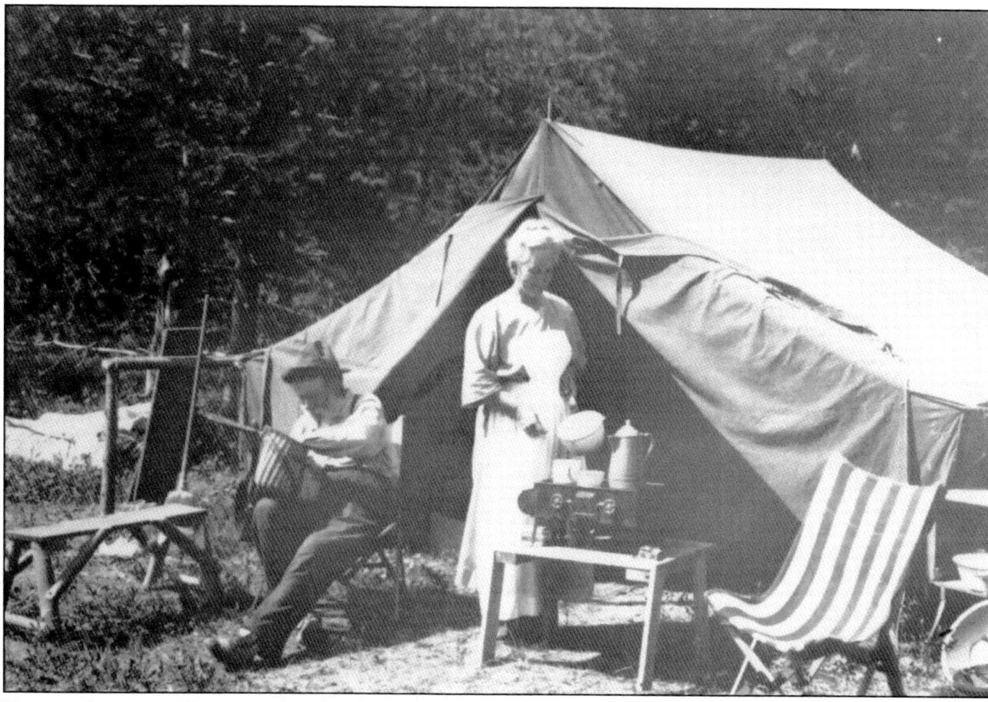

Family Camping Trip. It is hard to believe that tent camping in the mountains would be a relaxing activity for this family, given their rustic accommodations in a mining camp during the 1890s. These two photographs were taken in the Little Belt Mountains, around 1893. The man, woman, and two children are unidentified. The man is readying his creel, apparently heading out for some brookie in one of the mountain streams. The woman appears to be preparing lunch, and the kids are, of all things, digging holes and crawling in. As indicated by the packing materials beside them, Carnation and Del Monte were already shipping their products to the mine district. (Both, THM.)

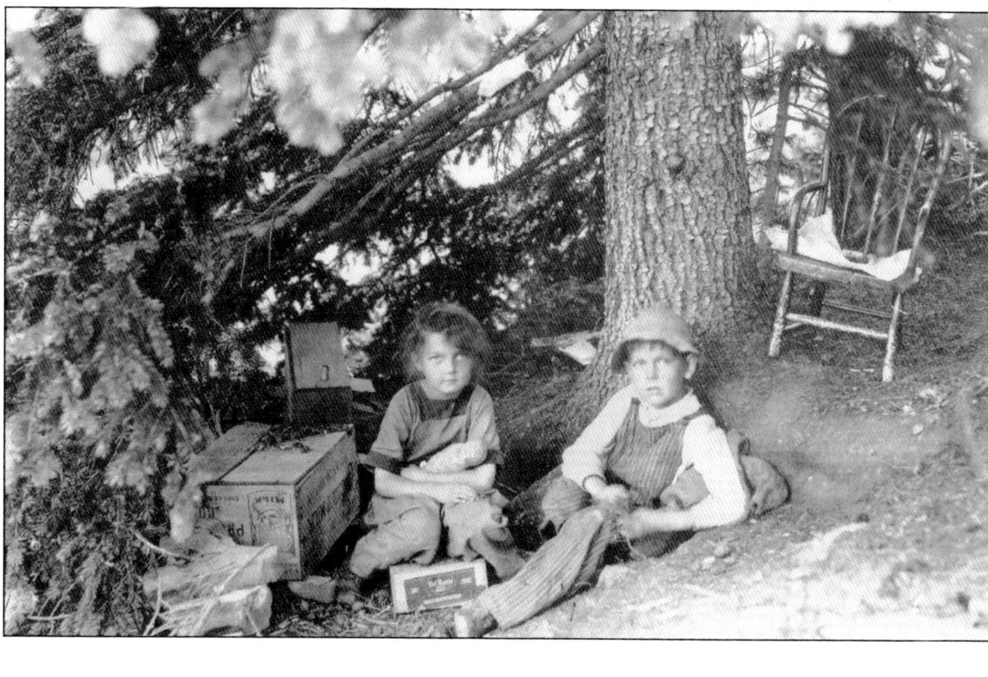

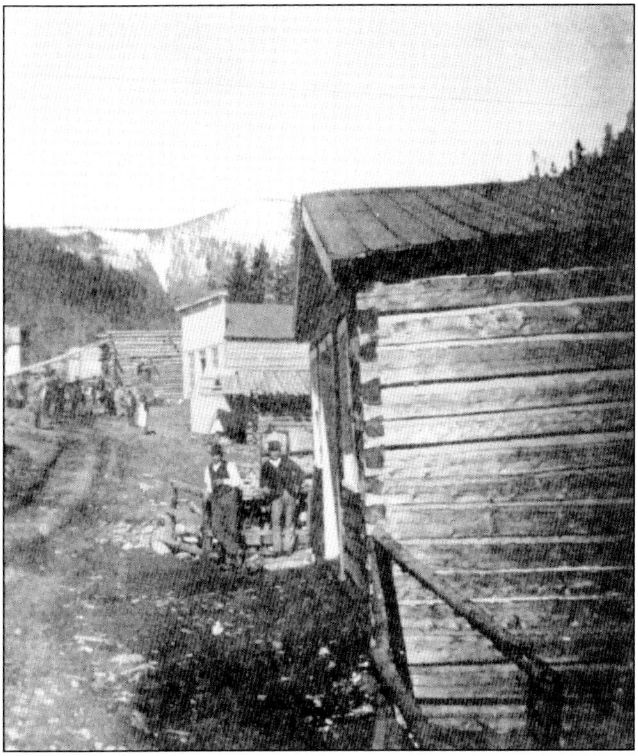

Barker Mining Camp. These photographs were taken by Daniel Dutro, the Neihart miner and mine owner who eventually became a photographer, producing a wonderful collection of historical images of early Montana. Dutro had been in Fort Benton before striking out, like so many others, to try his hand at prospecting. Following the sale of his mine interests (Benton Group), he returned to Fort Benton, purchased a photographic studio, and soon had a reputation as one of the finest photographers in the West. (MHS; photograph by Daniel Dutro.)

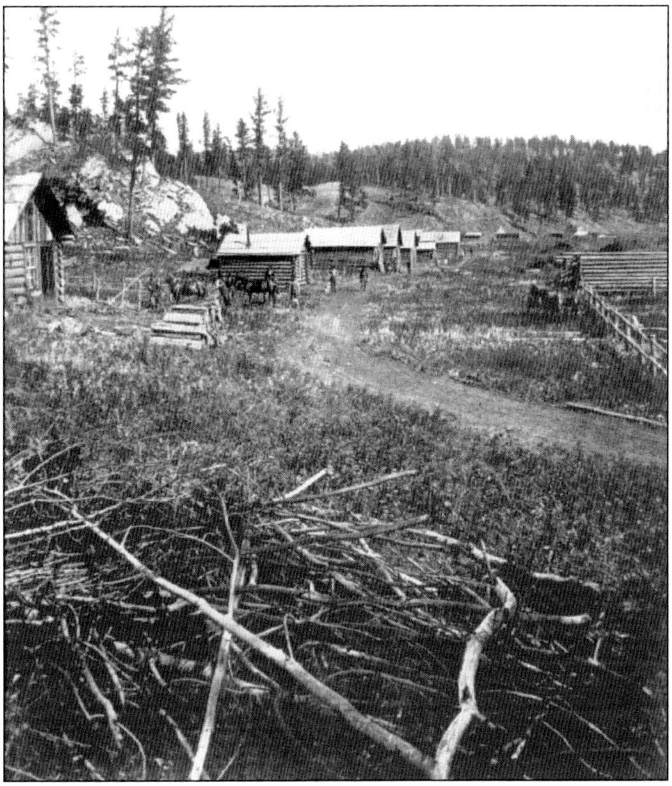

Another View of Barker Mining Camp. The two images on this page were taken in the mining district town of Barker and are from Dutro's "Frontier Sketches," housed in the archives of Montana's Historical Society in Helena. (MHS; photograph by Dan Dutro.)

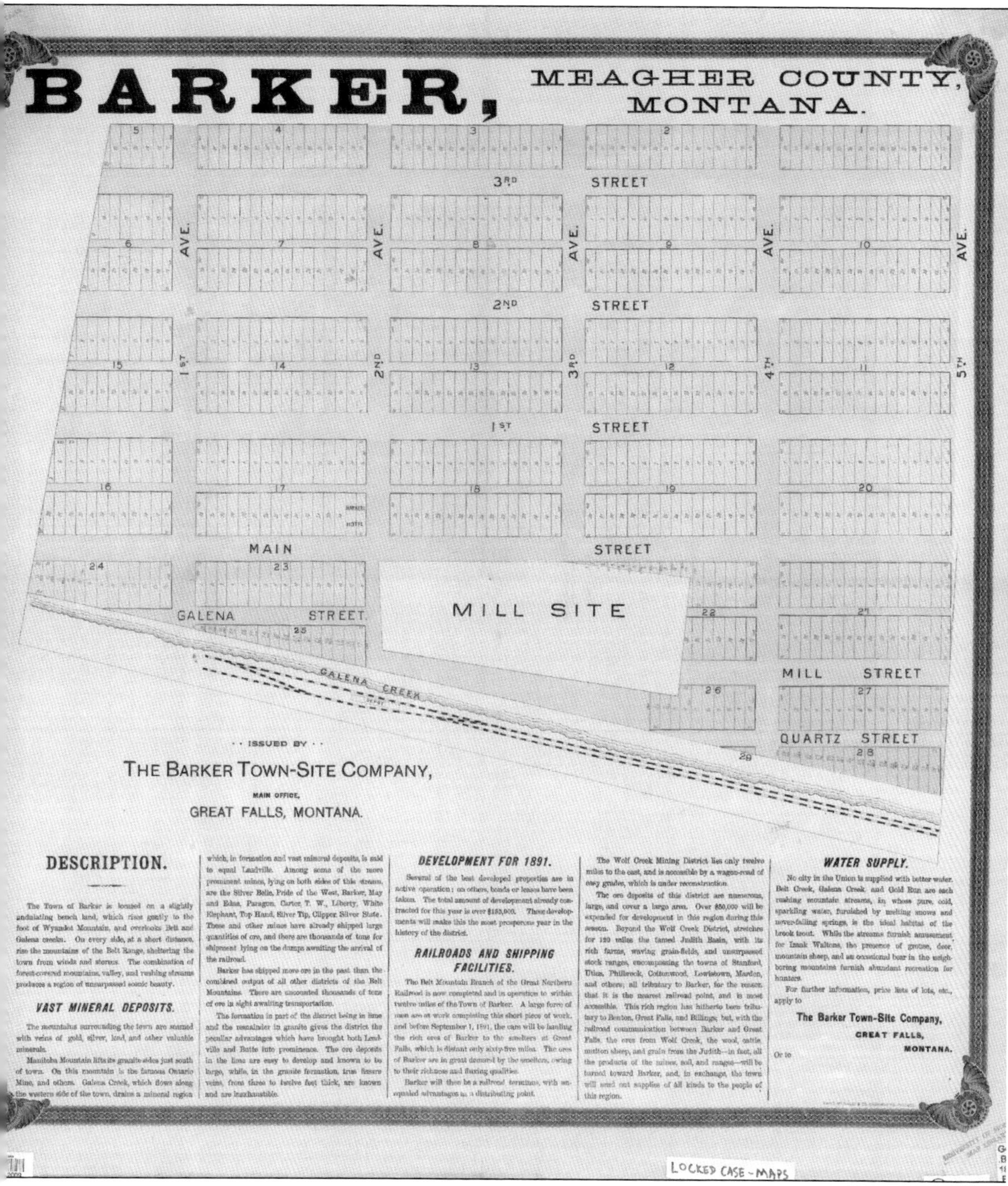

PLAT MAP OF BARKER, MONTANA. The theory was that hardrock prospectors and laborers followed a predictable path—first to Fort Benton on the Missouri River, then to Barker-Hughesville, and then to Neihart. Miners often were referred to as being "a product of (insert mining camp)." However, Barker's founders were not inclined to become a stopover for a traveler on his way to somewhere else. Seen here is the Barker Town-Site Company's plat map of 1895. (Archives and Special Collections, Mansfield Library, the University of Montana.)

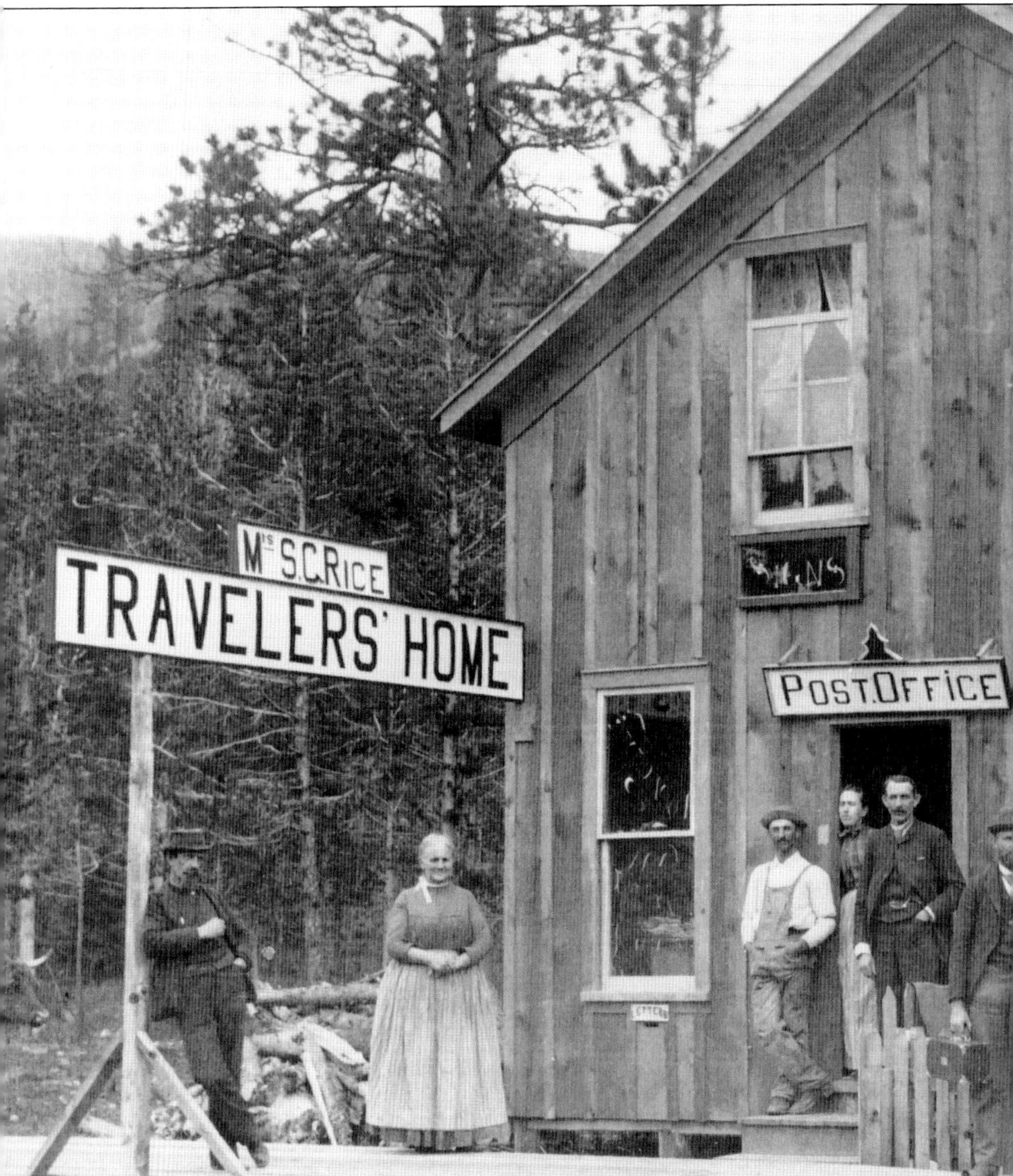

RICEVILLE POSTMISTRESS IN MONARCH. This photograph has caused some historians to scratch their collective heads. Mary E. Rice, pictured under the sign advertising her Travelers' Home (with her name prominently displayed on top), was appointed postmistress for Riceville on February 15, 1890. Monarch's first postmaster, Charles A. Martin, was appointed September 13, 1889, and Neihart's postmaster, Donald B. Mackintosh, was appointed August 21, 1882. However, it is believed that this site is located in Monarch, as a Dr. Craig is known to have practiced above

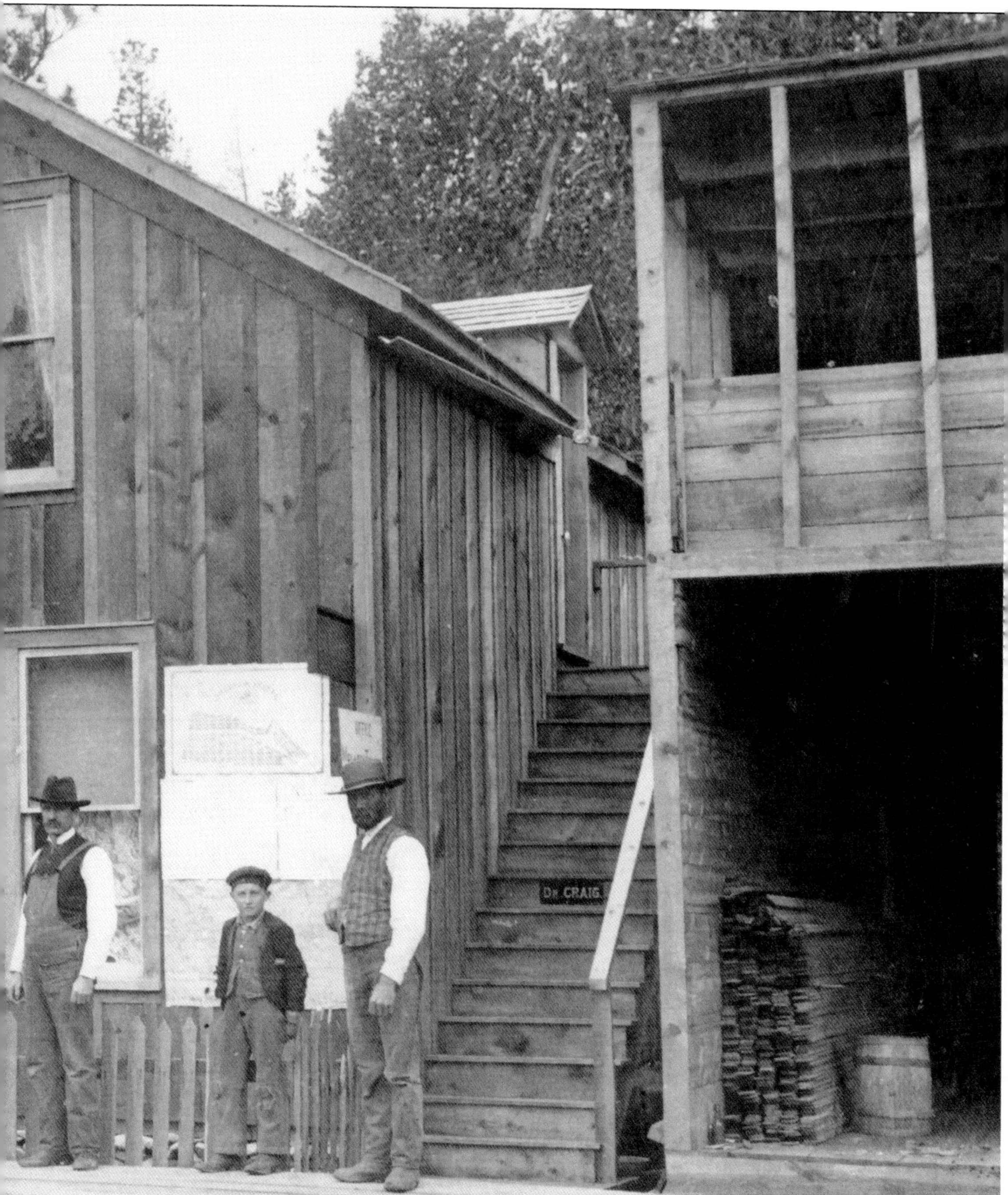

the Monarch Post Office in 1890, and his sign is visible in the image on one of the risers of the stairs leading up his office. The mystery that historians have wrestled with is the question of why Mary E. Rice would be credited as postmistress of Monarch. Perhaps she is only posing to advertise her Travelers' Home at that location in Monarch, not in her capacity as postmistress. The answer may never be certain. (MHS.)

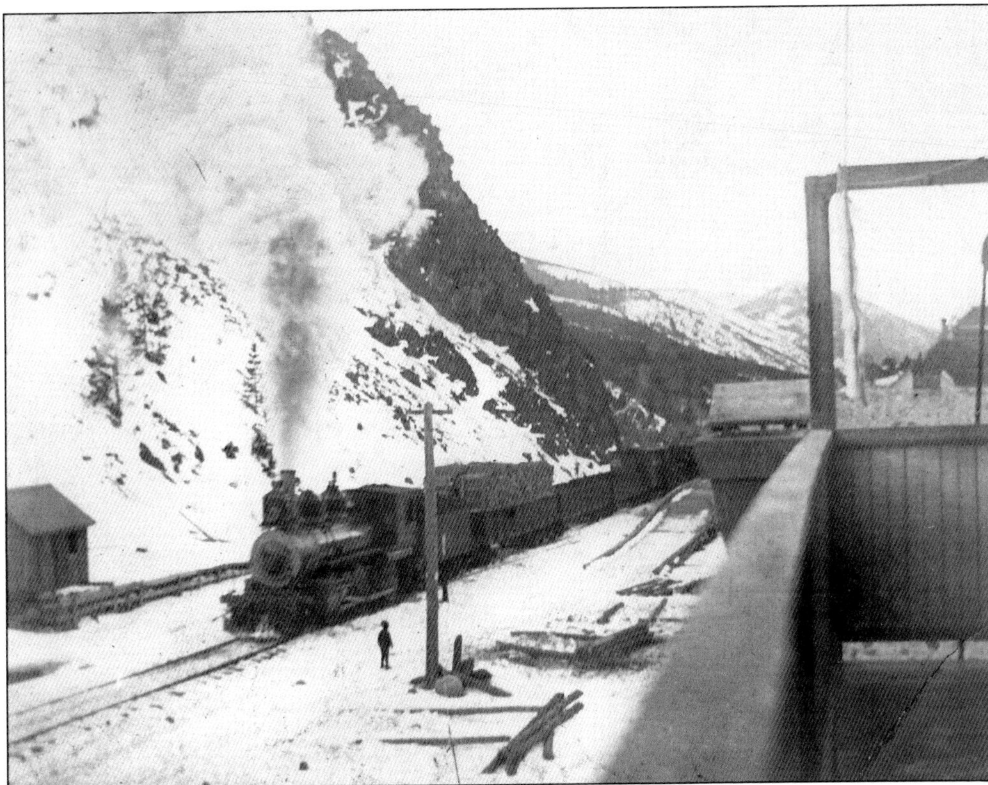

Steam Engines Provided Power for Freight. As these two photographs show, the power of steam locomotives allowed the shipment of iron ore and merchandise to and from the mines to continue in spite of poor weather. The train bellowing smoke passes close by one of the Neihart-area mines. The photograph below shows the Montana Central tracks (eventually part of the Great Northern Railroad) as they pass by Jericho, a suburb of Neihart. (Both, MHS.)

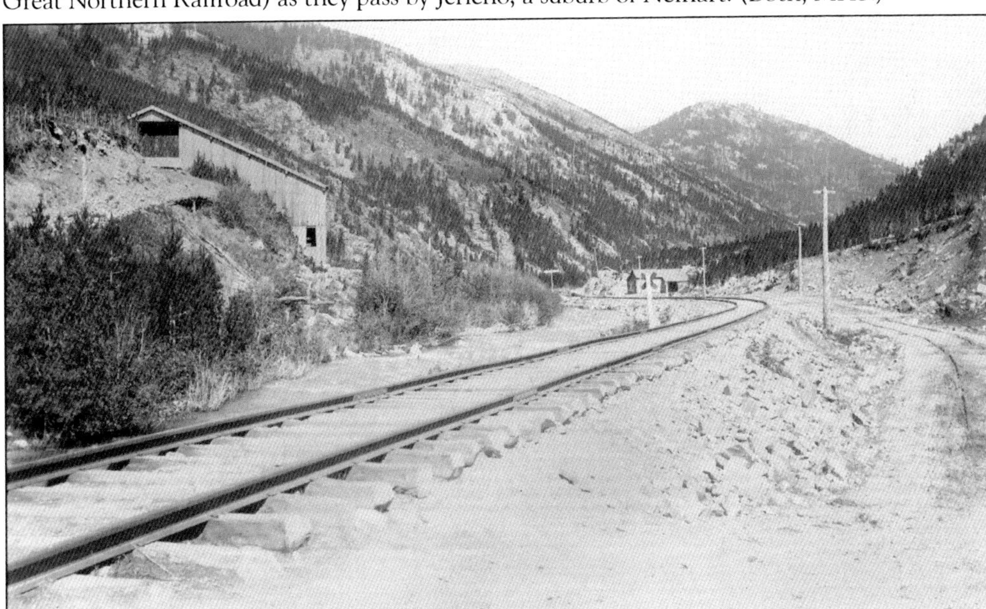

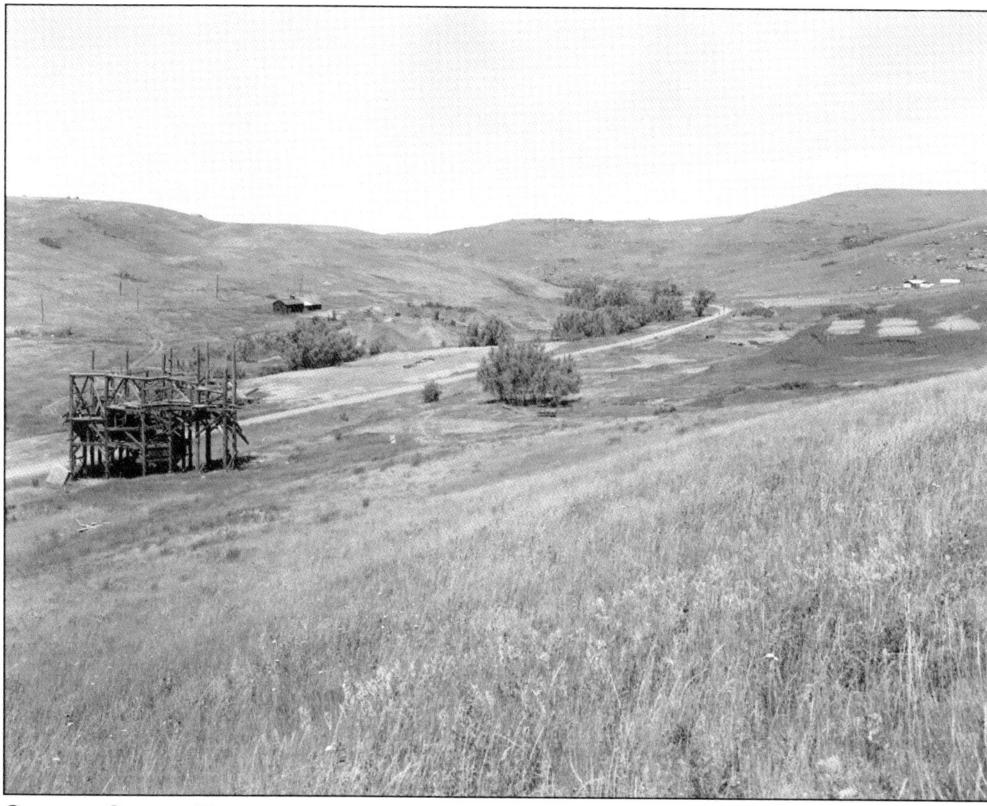

Cascade County Forgotten Mines. Not all mining activity in the Neihart, Monarch, and Barker-Hughesville areas was conducted in the Little Belt Mountains. These two photographs show abandoned structures related to the mining industry. Coal? Exploratory diggings? Unknown. What is known, however, is this area in Montana was active, productive, and successful due in large part to the determination and ingenuity of those early-era miners of the area. (Both, LOC.)

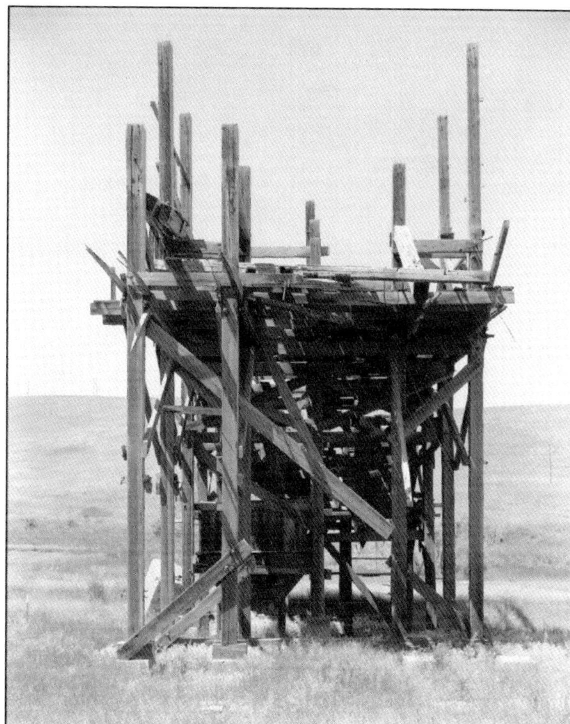

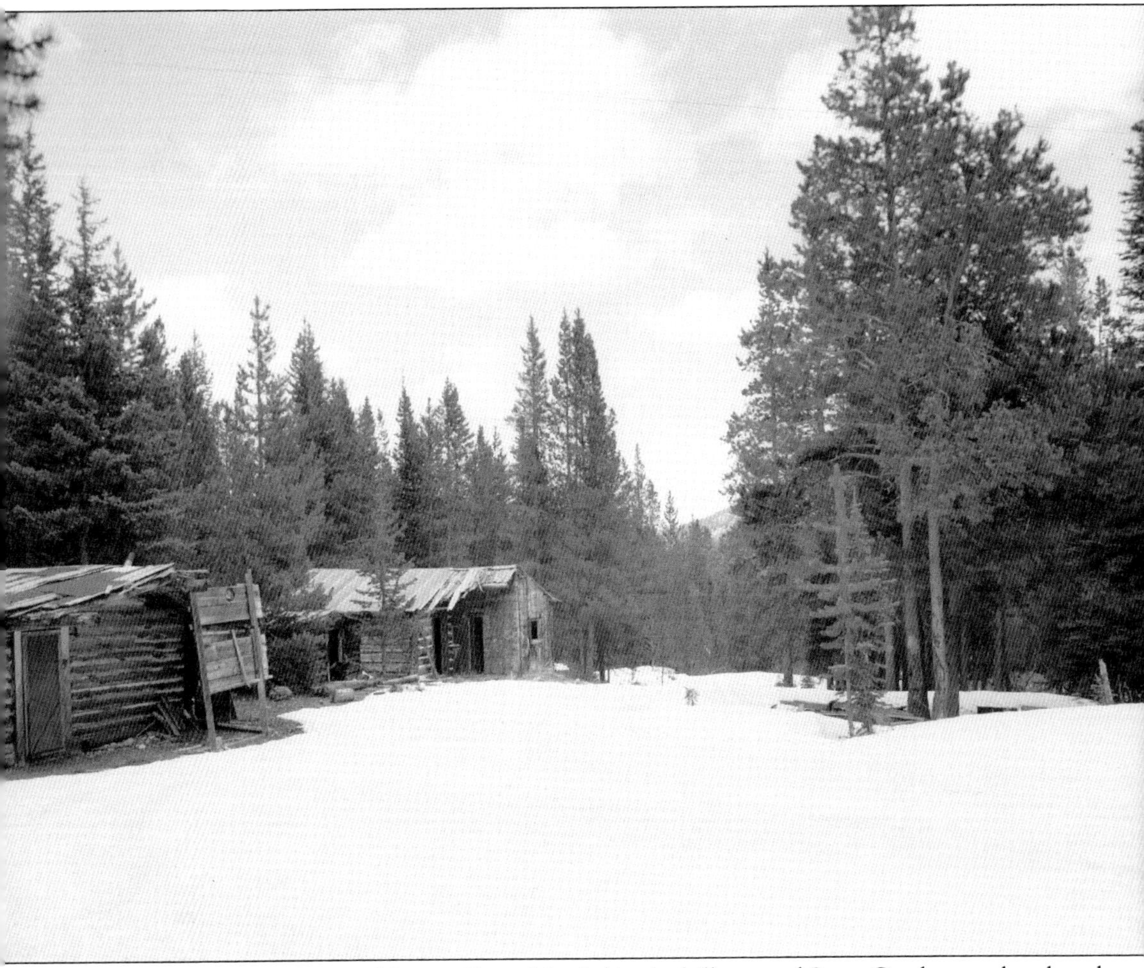

Winter View of Former Mining Camp. Nestled in the hills around Snow Creek were abandoned cabins, originally provided by the once-active Lexington Mine Company for its laborers. Miners, especially in remote locations like this, usually spent their winter months elsewhere, in places that did not subject them to the harsh realities of winter mining. (THM.)

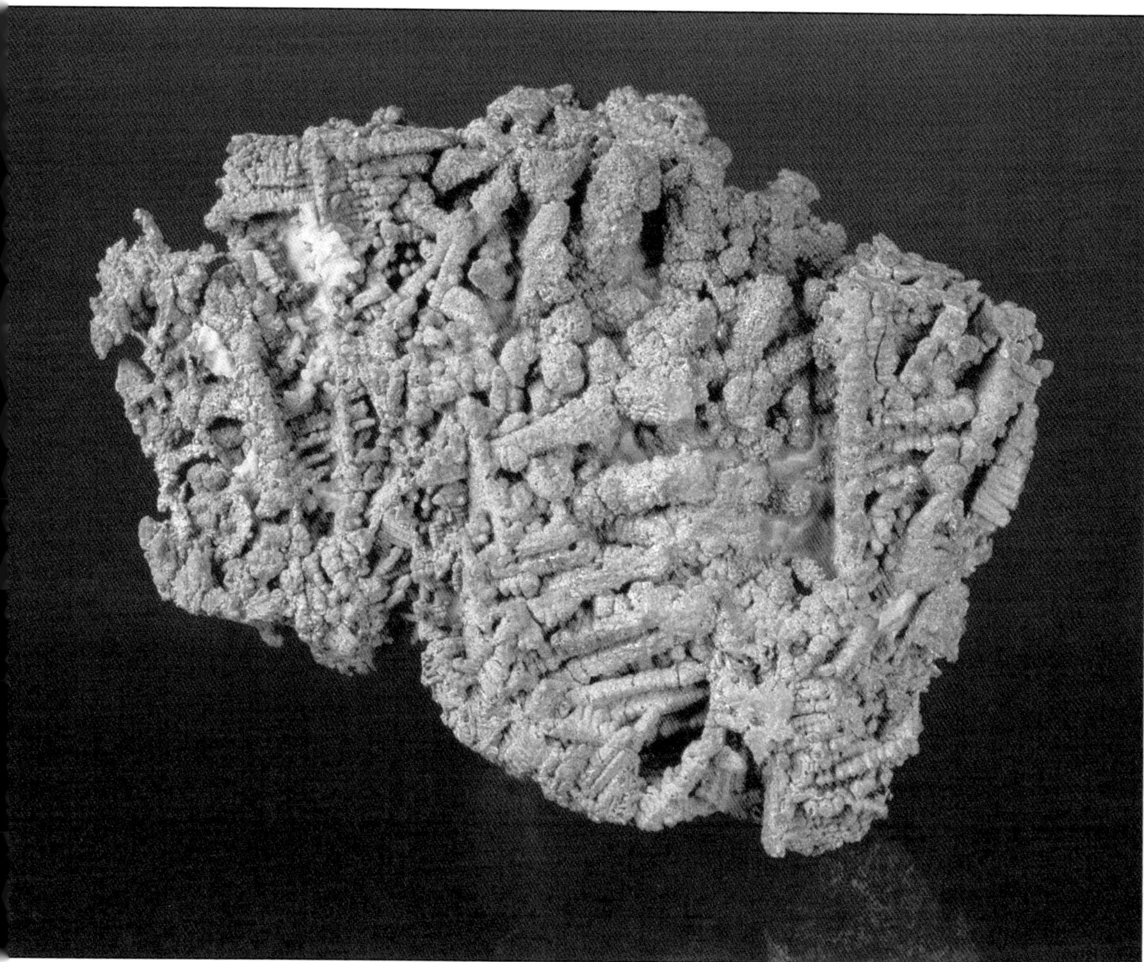

WHAT IT WAS ALL ABOUT.
Silver was discovered after gold and copper, about 4,000 BC, and it was used in jewelry and as a medium of exchange. Silver is generally found in the combined state in nature, usually in copper or lead mineralization, and by 2,000 BC, the mining and smelting of silver-bearing lead ores was under way. Lead ores were smelted to obtain an impure lead-silver alloy, which was then fire refined. In the mid-19th century a large silver deposit was discovered in Nevada. This, coupled with the silver deposits mined in Montana Territory and Idaho, resulted in the United States becoming the world's largest silver producer until the 20th century, when it was surpassed by Mexico and South America. (LOC.)

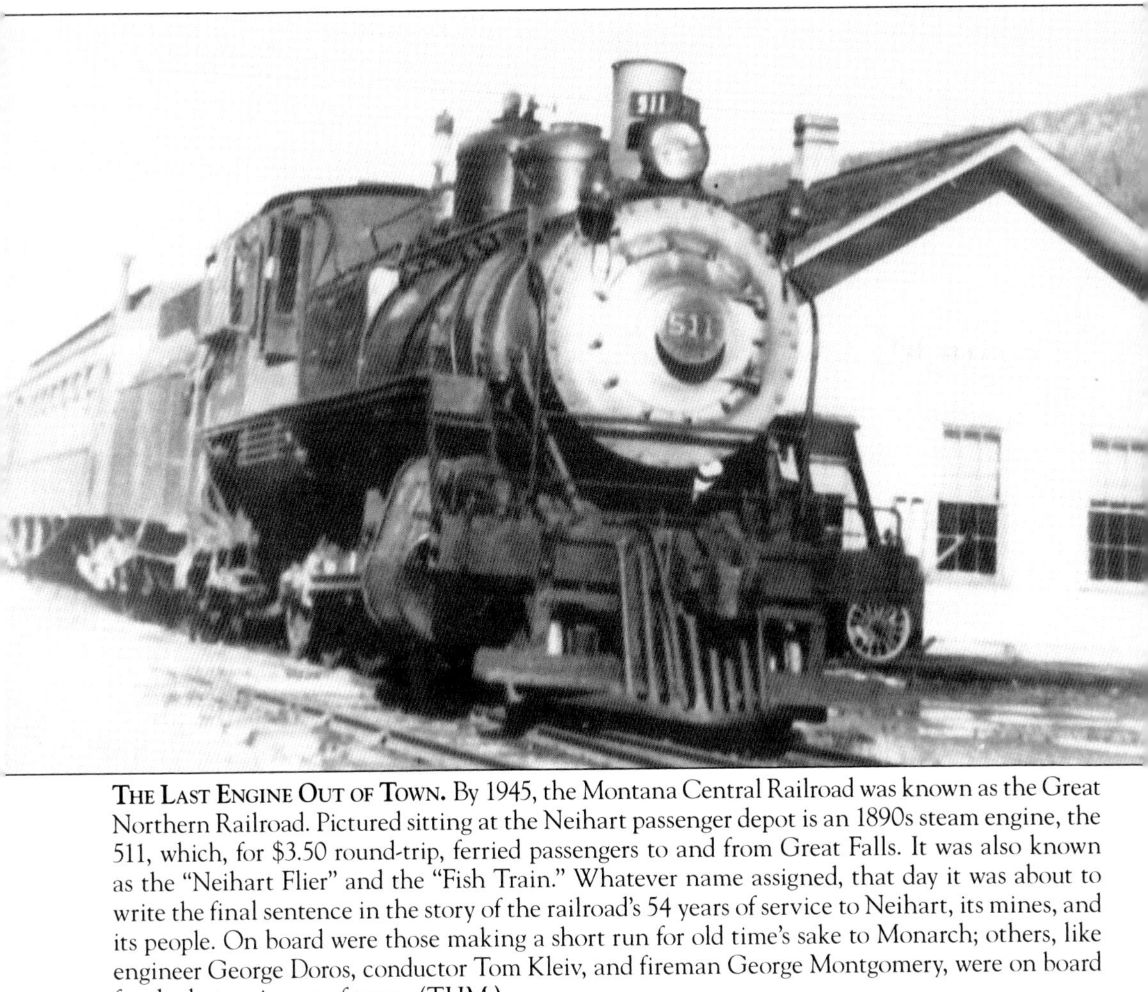

THE LAST ENGINE OUT OF TOWN. By 1945, the Montana Central Railroad was known as the Great Northern Railroad. Pictured sitting at the Neihart passenger depot is an 1890s steam engine, the 511, which, for $3.50 round-trip, ferried passengers to and from Great Falls. It was also known as the "Neihart Flier" and the "Fish Train." Whatever name assigned, that day it was about to write the final sentence in the story of the railroad's 54 years of service to Neihart, its mines, and its people. On board were those making a short run for old time's sake to Monarch; others, like engineer George Doros, conductor Tom Kleiv, and fireman George Montgomery, were on board for the last train out of town. (THM.)

Bibliography

Byrne, John, and Frank Hunter. *Twelfth Annual Report of the Inspector of Mines of the State of Montana*. Helena, MT: Independent Publishing Company, 1901.

Chadwick, Robert A. "Montana's Silver Mining Era: Great Boom and Great Bust." *Montana, the Magazine of Western History*, 32, No. 2 (Spring 1982): 16–31. (Accessed April 23, 2009, online.)

Chavez, Joel, compiled and edited. "Mining districts of Montana, Abandoned Mine Reclamation Bureau (AMRB)." Prepared by Montana State Library Natural Resource Information System (NRIS) for Montana Department of State Lands, Helena.

Leeson, M. A. (Michael A.). *History of Montana 1739–1885*. Chicago: Warner, Beers and Co., 1885.

Lutz, Dennis J. *Montana Post Offices & Postmasters*. Rochester, MN: Private printing, Johnson Printing Company, 1986.

MacDonald, Douglas, H. *Montana Before History 11,000 Years of Hunter-Gatherers in the Rockies and Plains*. Missoula, MT: Mountain Press Publishing Company, 2012.

Mining and Scientific Press. January 18, 1873, p. 23.

Malone, Michael P., Richard B. Roeder, and William L. Lang. *Montana: A History of Two Centuries*. Revised. Seattle: University of Washington Press, 1991.

Malone, Michael P. *James J. Hill Empire Builder of the Northwest*. The Oklahoma Western Biographies Series. Norman, OK: University of Oklahoma Press, 1996.

Neihart Herald Press. "The Herald Souvenir Thanksgiving." Illustrative. Reprinted. 1895.

Paul, Rodman, Wilson. *Mining Frontiers of the Far West 1848–1880*. Reprinted 1974 by University of New Mexico Press, original printing by Holt, Rinehart and Winston, Inc., 1963.

Perry, E. S. *Montana in the Geologic Past*. Reprinted 1986 by Montana Bureau of Mines and Geology, 1962.

Pielou, E.C. *After the Ice Age, The return of Life to Glaciated North America*. Chicago: University of Chicago Press, 1991.

Trimble, William J. "The Mining Advance into the Inland Empire: A Comparative Study of the Beginnings of the Mining Industry in Idaho and Montana, Eastern Washington and Oregon, and the southern Interior of British Columbia; and of Institutions and Laws Based upon that Industry." University of Wisconsin, Bulletin No. 638, History Series, Vol. III, No 2. 1914.

Wahlberg, Donna. *So Be It a History of the Barker Mining District Hughesville & Barker, Montana*. Self-published, 1989.

Weed, Walter Harvey, and Louis V. Pirsson. "Geology and Mineral Resources of the Judith Mountains of Montana." US Geological Survey, 18th Annual Report, Part 3, pp. 437–615. 1898.

Woodward, Lee A., and Otto L. Schumacher. *Sapphires, Gold, and Silver Field Guide to Little Belt Mountains, Montana*. Lewistown: Running Wolf Press, 2011.

Wolle, Muriel Sibell. *Montana Pay Dirt*. Athens, Ohio: Sage Books, 1963.

Young, Otis E. Jr. *Western Mining an Informal Account of Precious-Metals Prospecting, Placering, Lode Mining, and Milling from Spanish Times to 1893*. Norman, OK: University of Oklahoma Press, 1970.

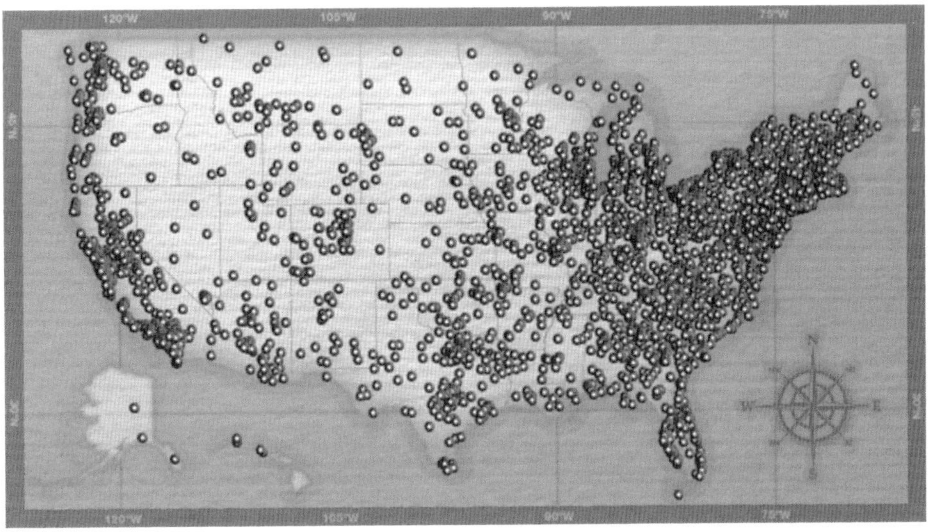